Springer Series in Microbiology

Editor: Mortimer P. Starr

Gerhard Gottschalk

Bacterial Metabolism

Springer-Verlag New York Heidelberg Berlin

Gerhard Gottschalk
Professor of Microbiology
Institut für Mikrobiologie der Universität Göttingen
34 Göttingen
Federal Republic of Germany

Series Editor:
Mortimer P. Starr
University of California
Department of Bacteriology
Davis, California 95616/USA

With 161 figures.

Library of Congress Cataloging in Publication Data

Gottschalk, Gerhard.
  Bacterial metabolism.
  (Springer series in microbiology)
  Includes bibliographical references and index.
  1. Microbial metabolism. 2. Bacteria—Physiology.
I. Title. DNLM: 1. Bacteria—Metabolism.
QW52.3 G687b
QR88.G67   589.9'01'33   78-7880

Printed in the United States of America.
9 8 7 6 5 4 3 2 1

ISBN 0-387-90308-9 Springer-Verlag New York
ISBN 3-540-90308-9 Springer-Verlag Berlin Heidelberg

To Ellen

Preface

This book has been written for students who are taking a course in bacterial metabolism. I hope, however, that scholars will also find it useful either as a help in teaching bacterial metabolism or as a review on the special aspects of metabolism in bacteria.

The concept of this book results from my experience in teaching bacterial metabolism. In the first chapters the principal reactions of the energy and biosynthetic metabolism have been discussed using *Escherichia coli* as a model organism. Then the diversity of aerobic metabolism has been outlined. Following a brief description of the regulation of the level and the activity of enzymes in bacteria the characteristic features of fermentative, chemolithotrophic and phototrophic metabolism have been discussed. Finally, the last chapter has been devoted to nitrogen fixation. Throughout the text I have tried not only to describe metabolic pathways and enzyme reactions but also to elucidate the physiology of the microorganisms which carry out all these metabolic reactions.

Two comments regarding the formulas used in this book are necessary. Organic acids are usually called after the names of their salts which are shorter (formate for formic acid, pyruvate for pyruvic acid). However, in schemes and figures the formulas of the free acids are given. Furthermore, it should be pointed out that $NADH_2$ and $NADPH_2$ and not NADH and NADPH are used as abbreviations for reduced nicotinamide-adenine dinucleotide and reduced nicotinamide-adenine dinucleotide phosphate, respectively. This has been done as these compounds are two electron carriers and redox reactions involving these carriers are thus easier to formulate.

I am particularly indebted to Joan Macy, Lynne Quandt, Jan Andreesen and Peter Hillmer for reading the manuscript, for their criticisms and their suggestions, and I thank Ute Gnass for typing the manuscript and for her invaluable help with the indexing and with the preparation of the figures. Finally, I am grateful to the publishers for their patience, willing help, and cooperation.

Göttingen, 1978   GERHARD GOTTSCHALK

Contents

Chapter 1

Nutrition of Bacteria

Bacteria, like all other living organisms, require certain nutrients for growth. These nutrients must contain those chemical elements that are constituents of the cellular materials and that are necessary for the activity of enzyme and transport systems. In addition, the nutrients must provide the organisms with materials for the production of biologically utilizable energy.

I. Major and Minor Bio-Elements

Only a small number of the elements of the periodic system are required by organisms in relatively high concentrations ($>10^{-4}$ M). These ten major bio-elements and some of their functions are presented in Table 1.1. Carbon, oxygen, hydrogen, and nitrogen are the main constituents of the organic compounds occurring in organisms. Sulfur is required for the synthesis of the amino acids cysteine and methionine and of a number of coenzymes. Phosphorus is present in nucleic acids, phospholipids, teichoic acids, and in nucleotides such as ATP, GTP, NAD, and FAD. The remaining four major bio-elements are metal ions, which are required as cofactors for enzyme activity and as components of metal complexes. Most of the biologically active phosphate esters are, for instance, present in the cell as magnesium complexes. The phospholipoproteins of bacterial cell walls are also chelated with magnesium ions.

Exoenzymes such as amylases and proteases are calcium proteins, and calcium dipicolinate is an important component of endospores. Ferrous and ferric ions are present in redox carriers such as cytochromes and iron-sulfur proteins.

Table 1.1. The ten major bio-elements, their sources, and some of their functions in microorganisms

element	source	function in metabolism
C	organic compounds, CO_2	main constituents of cellular material
O	O_2, H_2O, organic compounds, CO_2	
H	H_2, H_2O, organic compounds	
N	NH_4^+, NO_3^-, N_2, organic compounds	
S	SO_4^{2-}, HS^-, S^0, $S_2O_3^{2-}$, organic sulfur compounds	constituent of cysteine, methionine, thiamin pyrophosphate, coenzyme A, biotin, and α-lipoic acid
P	HPO_4^{2-}	constituent of nucleic acids, phospholipids, and nucleotides
K	K^+	principal inorganic cation in the cell, cofactor of some enzymes
Mg	Mg^{2+}	cofactor of many enzymes (e.g., kinases); present in cell walls, membranes, and phosphate esters
Ca	Ca^{2+}	cofactor of enzymes; present in exoenzymes (amylases, proteases); Ca-dipicolinate is an important component of endospores
Fe	Fe^{2+}, Fe^{3+}	present in cytochromes, ferredoxins, and other iron-sulfur proteins; cofactor of enzymes (some dehydratases)

Besides these ten major bio-elements, organisms require a number of others in small amounts (Table 1.2). Zinc and manganese ions are essential for all microorganisms. Zinc is especially important because RNA and DNA polymerase are zinc-metalloproteins. Sodium chloride is required by halophilic microorganisms in high concentration. This, however, is the exception. Most microorganisms have little use for sodium and chloride ions. Specific functions can be assigned to the other metals listed in Table 1.2. Molybdoproteins play an important role in nitrogen metabolism and in formate oxidation. Xanthine dehydrogenase also contains molybdenum. Of the selenoproteins listed in Table 1.2, the glycine reductase contains selenium in the form of selenocysteine. Cobalt is required by all organisms that perform B_{12}-dependent reactions. Copper is present in a number of

Table 1.2. Minor bio-elements, their sources, and some of their functions in micro-organisms

element	source	function in metabolism
Zn	Zn^{2+}	present in alcohol dehydrogenase, alkaline phosphatase, aldolase, RNA and DNA polymerase
Mn	Mn^{2+}	present in bacterial superoxide dismutase; cofactor of some enzymes (PEP carboxykinase, re-citrate synthase)
Na Cl	Na^+ Cl^-	required by halophilic bacteria
Mo	MoO_4^{2-}	present in nitrate reductase, nitrogenase, and formate dehydrogenase
Se	SeO_3^{2-}	present in glycine reductase and formate dehydrogenase
Co	Co^{2+}	present in coenzyme B_{12}-containing enzymes (glutamate mutase, methylmalonyl-CoA mutase)
Cu	Cu^{2+}	present in cytochrome oxidase and oxygenases
W	WO_4^{2-}	present in some formate dehydrogenases
Ni	Ni^{2+}	present in urease; required for autotrophic growth of hydrogen-oxidizing bacteria

enzymes transferring electrons from substrates to oxygen. Finally, tungsten and nickel are needed by microorganisms in some rare cases.

In nature, most of the bio-elements occur as salts, and they are taken up by the organisms as cations and anions, respectively. A greater diversity of compounds utilized by microorganisms is only observed with respect to the first five elements of Table 1.1: sulfur, nitrogen, oxygen, hydrogen, and carbon.

Sulfur is normally taken up as sulfate, reduced to the level of sulfide, and then used for biosynthetic purposes. Certain groups of bacteria, however, depend on the availability of reduced sulfur compounds. Some methane bacteria grow only in the presence of hydrogen sulfide as sulfur source. Thiobacilli and a number of phototrophic bacteria require sulfide, elemental sulfur, or thiosulfate as electron donor.

Nitrogen is required in large quantities because it amounts to approximately 10% of the dry weight of bacteria. It occurs naturally in the form of ammonia, nitrate, nitrite, nitrogen-containing organic compounds, and molecular nitrogen. The preferred source of nitrogen is ammonia, which can be utilized by practically all microorganisms. Nitrate is also taken up and used by many microorganisms but not by all. Before it can be incorporated into organic substances it has to be reduced to ammonia. Nitrite is the product of the nitrate–nitrite respiration and of the metabolic activities of *Nitrosomonas* and related species. A number of organisms reduce it to ammonia or N_2. Alternatively, nitrite can be oxidized to nitrate by *Nitrobacter* species. Several bacteria are able to fix molecular nitrogen and to reduce it to ammonia. This capacity is found only in certain prokaryotes but not in eukaryotes. Finally, organic compounds serve as nitrogen sources for many microorganisms. Usually these compounds are degraded such that ammonia becomes available for biosyntheses.

Carbon, **hydrogen**, and **oxygen** can be utilized by bacteria in the form of organic and inorganic compounds. Among the inorganic compounds used are CO_2, H_2, H_2S, NH_3, H_2O, O_2, NO_3^-, and SO_4^{2-}. On earth, not a single organic compound formed by organisms is accumulated. This implies that all of them are degradable. Microorganisms play an important role in this degradation. Their versatility has led to the formulation of the "doctrine of microbial catabolic infallibility," meaning that every naturally occurring carbon compound is used by some microbe.

The metabolism of carbon-, hydrogen-, and oxygen-containing compounds is not only important because these elements are the main constituents of the cell. These compounds are important substrates for the energy production in microorganisms.

II. The Two Basic Mechanisms of ATP Synthesis

The principal carrier of biologically utilizable energy is adenosine-5'-triphosphate (ATP), and all energy-requiring processes in living cells are directly or indirectly coupled to the conversion of ATP to adenosine-5'-diphosphate (ADP) and inorganic phosphate (P_i):

$$\text{ATP} + \text{H}_2\text{O} \xrightarrow{\text{cellular processes}} \text{ADP} + \text{P}_i$$

ATP contains two phosphate bonds with a high free energy of hydrolysis. The bonds are often symbolized by the squiggle "$\sim$":

Because of the high-energy phosphoryl bonds, ATP is an excellent phosphorylating agent, and it is used as such in a large number of reactions by all organisms. At the expense of ATP, intermediates of cell metabolism are activated for further reactions, such as condensations, reductions, and cleavages. Glutamine, for instance, can be synthesized from glutamate and ammonia only if a phosphorylated intermediate is formed. The reaction is, therefore, connected with the formation of ADP and P_i from ATP:

$$\text{glutamate} + NH_3 + ATP \longrightarrow \text{glutamine} + ADP + P_i$$

The high potential of group transfer of the AMP and the ADP group is also taken advantage of in a number of reactions; amino acids are activated by their conversion into the corresponding AMP derivatives with ATP, and AMP is released in the formation of aminoacyl-transfer-RNA:

$$\text{amino acid} + ATP \longrightarrow \text{aminoacyl-AMP} + \text{pyrophosphate}$$
$$\text{aminoacyl-AMP} + \text{transfer-RNA} \longrightarrow \text{aminoacyl-transfer-RNA} + AMP$$

The enzyme adenylate kinase catalyzes the phosphorylation of AMP to ADP with ATP; pyrophosphate (PP_i) is hydrolyzed to inorganic phosphate by pyrophosphatase so that the end products of this reaction series are also ADP and P_i:

$$AMP + ATP \xrightarrow{\text{adenylate kinase}} 2\ ADP$$

$$PP_i + H_2O \xrightarrow{\text{pyrophosphatase}} 2\ P_i$$

ADP and P_i are thus the principal products of the energy expenditure in metabolism, and the generation of ATP from ADP and P_i is a vital process of all living organisms. There are two basic mechanisms of ATP generation: electron transport phosphorylation and substrate-level phosphorylation.

Electron transport phosphorylation refers to a mechanism in which the flow of electrons from donors with a negative redox potential to acceptors with a more positive redox potential is coupled to the synthesis of ATP from ADP and P_i. Systems in which electron transport phosphorylation occurs are the respiratory chains and the photosynthetic apparatus.

Substrate-level phosphorylation is the second mechanism of ATP generation. During the degradation of organic substrates a small number of intermediates is formed containing high-energy phosphoryl bonds. Intermediates of this kind are:

$$
\begin{array}{l}
\overset{\displaystyle O}{\overset{\displaystyle \|}{C}}\!-\!O \sim PO_3H_2 \\
|\\
HC\!-\!OH \qquad\qquad \text{1,3-bisphosphoglycerate}\\
|\\
CH_2\!-\!O\!-\!PO_3H_2
\end{array}
$$

$$\begin{array}{l} CH_2 \\ \| \\ C\!-\!O\sim PO_3H_2 \\ | \\ COOH \end{array} \qquad \text{phospho-enolpyruvate}$$

$$\begin{array}{l} O \\ \| \\ C\!-\!O\sim PO_3H_2 \\ | \\ CH_3 \end{array} \qquad \text{acetyl phosphate}$$

The further metabolism of such organic $\sim P$ compounds is coupled to the transfer of the phosphate group to ADP, and this kind of ATP synthesis is called substrate-level phosphorylation:

$$\begin{array}{c} H_2C = C\!-\!COOH \\ | \\ O\sim PO_3H_2 \end{array} + ADP \longrightarrow CH_3\!-\!CO\!-\!COOH + ATP$$

III.   Nutrients as Energy Sources

It has already been mentioned that the function of the nutrients is not only to provide the organisms with the bio-elements. Nutrients are also required as energy sources—as fuel for the production of ATP. Various energy sources are available in nature and are taken advantage of by microorganisms, but they cannot be used by every bacterium, and it has become useful to group bacteria on the basis of their characteristic energy source. Organisms using light as energy source are called **phototrophs** (Greek phos=light, trophe= nutrition). If ATP comes from chemical reactions, the organisms which carry out this type of energy metabolism are called **chemotrophs.**

A.   Phototrophy

Phototrophs contain a photosynthetic apparatus that enables them to convert light energy into the high-energy phosphate bonds of ATP:

$$ADP + P_i \xrightarrow{\;light\;} ATP + H_2O$$

The carbon source commonly used by phototrophs is CO_2, and organisms that derive most of their cell carbon from CO_2 are called **C-autotrophs** (Greek autos=self; autotroph=self-feeding). Thus, phototrophic bacteria are C-autotrophic organisms, and when growing with CO_2, they require an electron donor for the reduction of CO_2 to the level of cell material. It is apparent from Table 1.3 that, as in plants and in blue-green bacteria, the electron donors used are frequently inorganic compounds. Phototrophic bacteria employ molecular hydrogen or reduced sulfur compounds, and

Table 1.3. The two types of phototrophic metabolism

type	electron donor	carbon source	examples
photolithotrophy	H_2O	CO_2	plants, blue-green bacteria
	H_2S, S^0, H_2	CO_2	*Chromatiaceae,* *Chlorobiaceae*
photoorganotrophy	organic substrates	organic substrates	*Rhodospirillaceae*

when doing so they are called **photolithotrophs** (Greek lithos = stone). A number of phototrophic bacteria can also grow in the light with organic substrates such as succinate or acetate. Under these conditions the source of any electrons used in reduction reactions is an organic substrate, and the bacteria grow then as **photoorganotrophs**. Clearly, the main source of cell carbon is then the organic substrate and not CO_2, and the organisms grow as **C-heterotrophs** (Greek heteros = the other; heterotroph = feeding on others). It follows that the terms C-autotroph and C-heterotroph reflect the nature of the carbon source, whereas the terms lithotroph and organotroph describe the nature of the electron donor used.

B. Chemotrophy

Most bacteria gain ATP by chemical reactions. These are commonly oxidation-reduction reactions, which means that one substrate is reduced at the expense of a second one:

$$X_{red} + A_{ox} \longrightarrow X_{ox} + A_{red}$$
$$ADP + P_i \qquad ATP + H_2O$$

Higher organisms can only use organic substrates as electron donors (X_{red}) and oxygen as electron acceptor (A_{ox}) and it is a specialty of the bacterial energy metabolism that, alternatively, other donors and acceptors can be employed. Here, A_{ox} may stand for oxygen, nitrate, sulfate, CO_2, or an organic compound, and X_{red} for an inorganic or an organic compound. By analogy to the nutritional classification of the phototrophs, bacteria which employ an organic compound as electron donor are called **chemoorganotrophs**. This group includes aerobes and anaerobes. The anaerobes, as is apparent from Table 1.4, use either nitrate, sulfate, or organic substrates as electron acceptors. Thus, organisms carrying out fermentations—such as the clostridia and lactic acid bacteria—belong to this group.

Chemolithotrophs use inorganic electron donors such as hydrogen,

Table 1.4. The two types of chemotrophic metabolism

type	electron donor	electron acceptor	carbon source	examples
chemoorgano-trophy	organic substrate	O_2	organic substrate	pseudomonads, bacilli
	organic substrate	NO_3^-	organic substrate	*Bacillus licheni-formis*
	organic substrate	SO_4^{2-}	organic substrate	sulfate reducers
	organic substrate	organic substrate	organic substrate	clostridia, lactic acid bacteria
chemolitho-trophy	H_2	O_2	CO_2	hydrogen-oxidizing bacteria
	H_2S	O_2	CO_2	thiobacilli
	H_2S	NO_3^-	CO_2	*Th. denitrificans*
	Fe^{2+}	O_2	CO_2	*Th. ferrooxidans*
	NH_3	O_2	CO_2	*Nitrosomonas*
	NO_2^-	O_2	CO_2	*Nitrobacter*
	H_2	CO_2	CO_2	methanogenic bacteria
	H_2	CO_2	CO_2	*Acetobacterium*

hydrogen sulfide, ferrous ions, nitrite, or ammonia. These compounds are oxidized with oxygen to water, sulfate, ferric ions, and nitrate, respectively, and these exergonic reactions are coupled to the formation of ATP from ADP and P_i:

$$H_2 + \tfrac{1}{2}O_2 \longrightarrow H_2O$$
$$ADP + P_i \qquad ATP + H_2O$$

Some organisms, e.g., *Thiobacillus denitrificans*, can replace oxygen by nitrate.

Although the methanogenic and acetogenic bacteria are quite different from all other chemolithotrophs, because they are strictly anaerobic organisms, they belong to this nutritional group. They gain ATP by reduction of CO_2 to methane or acetate with molecular hydrogen. Thus, only inorganic substrates are used for energy production.

Clearly, chemolithotrophs gain ATP without metabolizing an organic compound. Their carbon source is usually CO_2, and they are, therefore, C-autotrophs. However, many chemolithotrophs are not restricted to this kind of energy metabolism. Hydrogen-oxidizing bacteria, for instance, can grow as chemoorganotrophs (aerobically with carbohydrates or organic acids) under appropriate conditions. They are, therefore, called facultative chemolithotrophs (all hydrogen-oxidizing bacteria, some thiobacilli). Species (*Nitrosomonas*, *Thiobacillus thiooxidans*), which are unable to grow in the absence of their inorganic electron donors, are called obligate chemolithotrophs.

IV. Growth Factor Requirements of Bacteria

So far it has been presumed that microorganisms themselves are able to synthesize all organic compounds required for growth. In fact, there are C-autotrophic bacteria that derive their cell carbon from CO_2 exclusively (e.g., *Alcaligenes eutrophus* and *Nitrobacter winogradskyi*) and there are also C-heterotrophs that grow on simple carbon sources such as glucose (e.g., *Escherichia coli*, *Bacillus megaterium*, and *Clostridium pasteurianum*). However, a number of bacteria lack the ability to synthesize all the organic compounds needed for growth and depend on certain growth factors. These factors can be combined to form three groups:

1. vitamins and related compounds, required in small amounts
2. amino acids
3. purines and pyrimidines

Number and kind of growth factors, which must be present in the growth medium, differ among bacteria. Lactic acid bacteria require practically all amino acids, purines, pyrimidines, and vitamins for growth. Their biosynthetic capacity is rather limited. Common among microorganisms are requirements for vitamins and related compounds. Some of them and their functions in metabolism are summarized in Table 1.5.

The exact growth factor requirements are not known for all microorganisms, and microbiologists add yeast extract and peptone to the growth media as complex and cheap sources of these factors. Synthetic media— media of known composition—are used for special purposes provided the requirements of a particular organism are known. *Clostridium kluyveri* grows in a medium supplemented with biotin and *p*-aminobenzoic acid. To the media for phototrophic bacteria, a vitamin solution is added containing nicotinic acid, thiamin, *p*-aminobenzoic acid, biotin, and vitamin B_{12}. Some organisms exhibit special requirements. A medium for *Haemophilus* species must contain hemin for cytochrome biosynthesis and also NAD. *Bacteroides* species require hemin. *Methanobacterium ruminantium* grows only if coenzyme M (2-mercaptoethanesulfonic acid) and 2-methyl-*n*-butyric acid are present. These few examples document that microoganisms may exhibit various defects in their biosynthetic machinery, and that growth factors are important for many of them.

Table 1.5. Vitamins and related compounds and their functions in metabolism

compound	function in metabolism
p-aminobenzoic acid	precursor of tetrahydrofolic acid, a coenzyme involved in transfer of one-carbon units
biotin	prosthetic group of enzymes catalyzing carboxylation reactions
coenzyme M	coenzyme involved in methane formation
folic acid	tetrahydrofolic acid is a coenzyme involved in transfer of one-carbon units
hemin	precursor of cytochromes
lipoic acid (dithiooctanic acid)	prosthetic group of the pyruvate dehydrogenase complex
nicotinic acid	precursor of NAD and NADP, which are coenzymes of many dehydrogenases
pantothenic acid	precursor of coenzyme A and of the prosthetic group of acyl carrier proteins
pyridoxine (vitamin B_6)	pyridoxal phosphate is a coenzyme for transaminases and amino acid decarboxylases
riboflavin (vitamin B_2)	precursor of flavin mononucleotide (FMN) and flavin adenine dinucleotide (FAD), which are the prosthetic groups of flavoproteins
thiamin (vitamin B_1)	thiamin pyrophosphate is the prosthetic group of decarboxylases, transaldolases, and transketolases
vitamin B_{12} (cyanocobalamin)	coenzyme B_{12} is involved in rearrangement reactions (e.g., glutamate mutase)
vitamin K	precursor of menaquinone, which functions as electron carrier (e.g., fumarate reductase)

V. Summary

1. Ten chemical elements are required by organisms in relatively high concentrations: C, O, H, N, S, P, K, Mg, Ca, Fe.

2. The minor bio-elements comprise some that are essential for all micro-organisms (Zn, Mn) and others that are required only in connection with special metabolic activities (e.g., Se, Mo, Co, Cu, W).

3. ATP is synthesized from ADP and P_i either by electron transport phosphorylation or by substrate-level phosphorylation. The energy for ATP synthesis is provided either as physical (light) or as chemical energy.

4. Phototrophic bacteria, which use inorganic electron donors such as H_2 or H_2S for the reduction of CO_2 to cell carbon, are called photolitho-trophs. Organisms that grow on organic substrates in the light are called photoorganotrophs.

5. Chemotrophic organisms derive energy from chemical reactions, in most cases from oxidation reactions. Chemoorganotrophs metabolize organic substrates. If only inorganic compounds are involved in energy production, the organisms are called chemolithotrophs.

6. With respect to the origin of the cell carbon, C-heterotrophs and C-autotrophs are distinguished, the former use organic compounds and the latter CO_2 as the main carbon source.

7. In addition to their simple carbon sources, many microorganisms require one or several growth factors for growth. These factors are vitamins and related compounds, amino acids, and purines and pyrimidines. Very common is a requirement for vitamins such as biotin, p-aminobenzoic acid, thiamin, nicotinic acid, and vitamin B_{12}.

Chapter 2

How *Escherichia coli* Synthesizes ATP during Aerobic Growth on Glucose

Escherichia coli belongs to the group of facultatively anaerobic bacteria. It is able to grow with a number of substrates in the presence of oxygen or in its absence. Under aerobic conditions part of the substrate is oxidized to CO_2 with oxygen as the terminal electron acceptor. This process is exergonic and allows for the formation of ATP, which is required for the biosynthesis of cellular constituents.

If glucose is the substrate about 50% is oxidized to CO_2; this results in enough ATP to convert the other 50% into cell material:

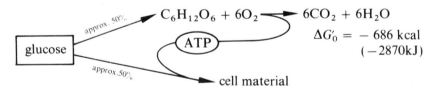

$$C_6H_{12}O_6 + 6O_2 \longrightarrow 6CO_2 + 6H_2O$$
$$\Delta G_0' = -686 \text{ kcal}$$
$$(-2870 \text{kJ})$$

In order to utilize the energy released during glucose oxidation effectively for the formation of ATP from ADP and inorganic phosphate, the glucose molecule must undergo a series of reactions, which at first sight appear fairly complicated but nevertheless are very economical. All the reactions involved in the oxidation of glucose to CO_2 can be divided into a number of functional blocks of reactions; in *E. coli* these are:

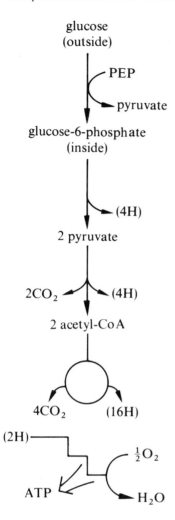

A. Transport of glucose into the cell by the phospho-enolpyruvate phosphotransferase system.

B. Degradation of glucose-6-phosphate to pyruvate via the Embden–Meyerhof–Parnas pathway.

C. Oxidative decarboxylation of pyruvate to acetyl-coenzyme A by puruvate dehydrogenase.

D. Oxidation of the acetyl moiety of acetyl-coenzyme A to CO_2 via the tricarboxylic acid cycle.

E. Oxidation of the reduced coenzymes formed in steps B to D in the respiratory chain.

These reactions together accomplish the oxidation of glucose to CO_2 and water, with the conservation of part of the energy released as phosphate-bond energy of ATP.

I.   Transport of Glucose into the *E. coli* Cell

The cytoplasmic membrane of *E. coli* is not simply permeable to glucose; there is no free diffusion of glucose in and out of the bacterial cells which would allow the concentration of the sugar inside and outside to become equal. Instead *E. coli* possesses a transport system that recognizes glucose specifically. It "picks up" glucose at the medium side of the membrane and releases it on the cytoplasm side of the membrane. This transport process is coupled to a chemical conversion of the substrate, i.e., the phosphorylation of glucose to glucose-6-phosphate.

The phosphate donor in this reaction is phospho-enolpyruvate and the enzyme complex catalyzing this transport process is called **phospho-enolpyruvate:glucose phosphotransferase system**. It is composed of two reactions:

1. A small protein, designated HPr, is phosphorylated by phospho-enolpyruvate (PEP):

$$PEP + HPr \underset{\longleftarrow}{\overset{enzyme\ I}{\longrightarrow}} phospho\text{-}HPr + pyruvate$$

This reaction is not specific to the glucose transport system but is involved in sugar transport of *E. coli* in general. Enzyme I and HPr serve as catalysts in transport systems for glucose, mannose, fructose, and other hexoses; both proteins are soluble.

2. The phosphoryl group of phospho-HPr is transferred to glucose and the glucose-6-phosphate is released into the cytoplasm. This complex reaction is catalyzed by enzyme II, which is specific for glucose and is membrane-bound:

$$phospho\text{-}HPr + glucose \xrightarrow{enzyme\ II} glucose\text{-}6\text{-}phosphate + HPr$$

Figure 2.1 summarizes both reactions. Transport processes, which are coupled to a conversion of the substrate (glucose) into a derivative of the substrate (glucose-6-phosphate), are called **translocation** processes.

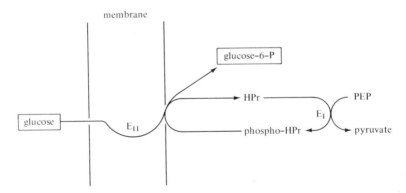

Figure 2.1. Transport of glucose by the phospho-enolpyruvate:glucose phosphotransferase system. Note that free glucose does not appear inside the cell. It is phosphorylated by the membrane-bound enzyme II and released into the cytoplasm as glucose-6-phosphate.

II.   Degradation of Glucose-6-Phosphate to Pyruvate via the Embden-Meyerhof-Parnas (EMP) Pathway

The EMP pathway was first discovered in muscle tissues. It is the most commonly used sequence of reactions for the conversion of glucose-6-phosphate into pyruvate and occurs in animals, plants, and many bacteria. *E. coli* contains high activities of the necessary enzymes. In the first two reactions catalyzed by the enzymes glucose phosphate isomerase and phosphofructokinase, glucose-6-phosphate is converted into **fructose-1,6-bis-phosphate**, the characteristic intermediate of this pathway. Fructose bisphosphate aldolase splits fructose-1,6-bisphosphate into two C_3 fragments— dihydroxyacetonephosphate and glyceraldehyde-3-phosphate. Both compounds are in equilibrium with each other due to the presence of the enzyme triosephosphate isomerase. The equilibrium constant of the isomerase reaction favors the dihydroxyacetonephosphate but the enzyme triosephosphate isomerase is so active (compared to other enzymes of the EMP pathway) that immediate conversion of the ketose derivative into the aldose isomer occurs. Consequently, if only little dihydroxyacetonephosphate is needed by the cells, most of the C_3 fragments originating from fructose-1,6-bisphosphate can be further metabolized via glyceraldehyde-3-phosphate. This is what normally happens (Figure 2.2).

The oxidation of glyceraldehyde-3-phosphate to pyruvate is initiated by two enzymes, which together accomplish the oxidation of the aldehyde group to a carboxyl group. The first of these is glyceraldehyde-3-phosphate dehydrogenase. It contains bound NAD, and the reaction proceeds as shown in Figure 2.3.

The 1,3-bisphosphoglycerate formed is a mixed anhydride of 3-phosphoglycerate and phosphate, and the free energy of hydrolysis of this anhydride is higher than that for $ATP \xrightarrow{\text{H}_2\text{O}} ADP + P_i$. Therefore, the conversion of 1,3-bisphosphoglycerate to 3-phosphoglycerate can be coupled to the phosphorylation of ADP to ATP. The enzyme catalyzing this reaction is 3-phosphoglycerate kinase. This is one of the two sites of the EMP pathway that yield ATP by substrate-level phosphorylation:

$$\text{1,3-bisphosphoglycerate} + \text{ADP} \xrightarrow{\text{3-phosphoglycerate kinase}} \text{3-phosphoglycerate} + \text{ATP}$$

In the next two steps 3-phosphoglycerate is converted to phospho-enolpyruvate. First phosphoglycerate mutase transfers the phosphoryl group from position three to position two of glycerate. The enzyme requires 2,3-bisphosphoglycerate as cofactor. Enolase (phosphopyruvate hydratase) removes water to yield phospho-enolpyruvate. This is another compound containing a phosphoryl bond with a high free energy of hydrolysis, and during the formation of pyruvate from phospho-enolpyruvate it is transferred to ADP to yield ATP. The enzyme catalyzing this reaction is pyruvate kinase

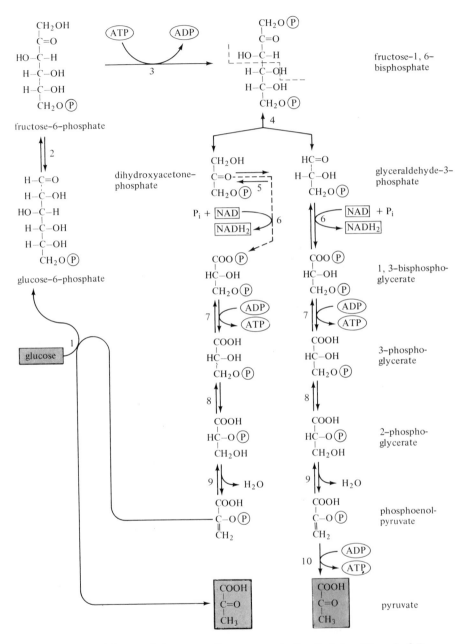

Figure 2.2. Breakdown of glucose to 2 pyruvate via the Embden–Meyerhof–Parnas pathway. 1, PEP:glucose phosphotransferase; 2, glucose phosphate isomerase; 3, phosphofructokinase; 4, fructose bisphosphate aldolase; 5, triose phosphate isomerase; 6, glyceraldehyde-3-phosphate dehydrogenase; 7, 3-phosphoglycerate kinase; 8, phosphoglycerate mutase; 9, enolase; 10, pyruvate kinase.

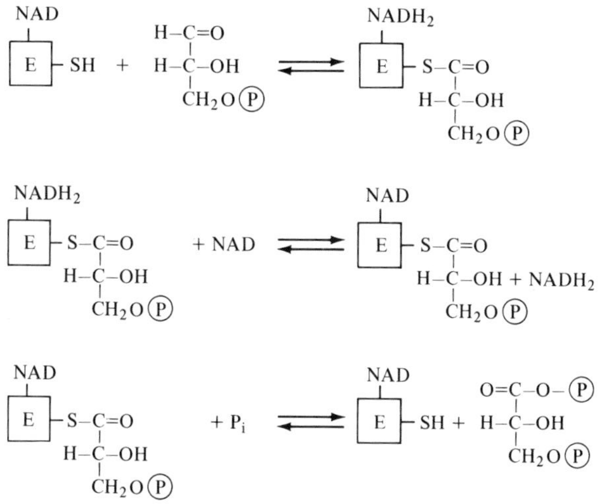

Figure 2.3. The conversion of glyceraldehyde-3-phosphate to 1,3-bisphosphoglycerate. In the initial reaction the aldehyde is oxidized to a thioester and the reducing power is transferred to the enzyme-bound NAD. An exchange reaction then takes place with soluble NAD. Finally, the acyl group on the enzyme is transferred to inorganic phosphate.

(second site of ATP formation in the EMP pathway):

$$\text{PEP} + \text{ADP} \xrightarrow{\text{pyruvate kinase}} \text{pyruvate} + \text{ATP}$$

Since phospho-enolpyruvate is required for the glucose transport system, only half of it is available for the pyruvate kinase reaction.

The sum of the reactions discussed thus far is:

$$\text{glucose} + \text{PEP} \longrightarrow \text{glucose-6-phosphate} + \text{pyruvate}$$
$$\text{glucose-6-phosphate} + \text{ADP} + 2\text{NAD} + 2\text{P}_i \longrightarrow 2\text{PEP} + \text{ATP} + 2\text{NADH}_2$$
$$\text{PEP} + \text{ADP} \longrightarrow \text{pyruvate} + \text{ATP}$$

$$\text{glucose} + 2\text{ADP} + 2\text{NAD} + 2\text{P}_i \longrightarrow 2 \text{ pyruvate} + 2\text{ATP} + 2\text{NADH}_2$$

NADH_2 is formed in the glyceraldehyde-3-phosphate dehydrogenase reaction and ATP by 3-phosphoglycerate kinase and pyruvate kinase. One ATP is consumed in the phosphofructokinase reaction.

III.   Oxidative Decarboxylation of Pyruvate to Acetyl-CoA

As in most aerobic microorganisms the formation of acetyl-CoA from pyruvate in *E. coli* is catalyzed by the **pyruvate dehydrogenase complex**. This complex consists of three enzymes: 24 molecules each of pyruvate dehydrogenase (E_1) and dihydrolipoic transacetylase (E_2) and 12 molecules of dihydrolipoic dehydrogenase (E_3). E_1 contains thiamin pyrophosphate (TPP), and the first step in the oxidative decarboxylation is the addition of pyruvate to C-2 of the thiazolium ring of TPP to form lactyl-TPP-E_1.

$$H_3C-\overset{+}{C}-C\overset{O}{\underset{O^-}{\diagdown}}$$

The further reactions acting upon the lactyl residue are summarized in Figure 2.4. Decarboxylation yields hydroxyethyl-TPP-E_1. The hydroxyethyl moiety is then transferred from TPP to the lipoate group of E_2. Concomitantly the disulfide bond of lipoate is reduced. The acetyl group thus formed is then released as acetyl-CoA and under catalysis of E_3 the sulfhydryl form of lipoate is oxidized by NAD to the disulfide form. The enzyme complex is then ready for the oxidation of another molecule of pyruvate to acetyl-coenzyme A.

The combined action of the enzymes of the EMP pathway and the

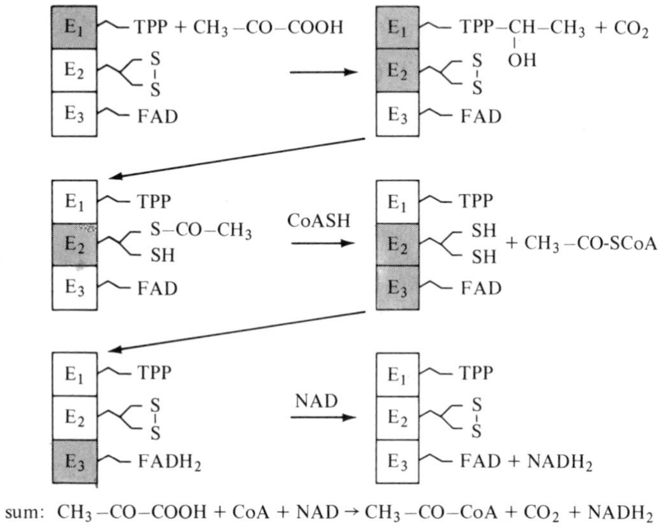
E1
TPP + CH3–CO–COOH
E2
S
S
E3
FAD
TPP–CH–CH3 + CO2
OH
TPP
S–CO–CH3
SH
CoASH
SH
SH
+ CH3–CO-SCoA
NAD
FADH2
FAD + NADH2
sum: CH3–CO–COOH + CoA + NAD → CH3–CO–CoA + CO2 + NADH2

Figure 2.4. Reactions catalyzed by the pyruvate dehydrogenase complex. E_1, pyruvate dehydrogenase; E_2, dihydrolipoic transacetylase; E_3, dihydrolipoic dehydrogenase; TPP, thiamin pyrophosphate; the disulfide compound linked to E_2 is the oxidized form of lipoate.

pyruvate dehydrogenase complex leads to the degradation of glucose to CO_2 and acetyl-CoA according to the equation:

$$\text{glucose} + 2\text{ADP} + 4\text{NAD} \xrightarrow{2\text{P}_i;\, 2\text{CoA}} 2\ \text{acetyl-CoA} + 2\text{CO}_2 + 2\text{ATP} + 4\text{NADH}_2$$

IV. Oxidation of Acetyl-CoA via the Tricarboxylic Acid Cycle

This cycle was discovered by Egleston and Krebs in animal tissues. It is often referred to as the Krebs or citric acid cycle. That it is present in *E. coli* has been demonstrated with enzymatic methods and with experiments using radioactive substrates. The cycle carries out the oxidation of the acetyl moiety of acetyl-CoA to CO_2 with transfer of the reducing equivalents to NAD, NADP, and FAD. Acetyl-CoA enters the cycle by the **citrate synthase** reaction in which oxaloacetate and acetyl-CoA are condensed to give citrate. In the subsequent reactions of the cycle the C_4-acceptor for the next molecule of acetyl-CoA, oxaloacetate, is rather elegantly regenerated.

As shown in Figure 2.5 citrate is first isomerized to isocitrate. This is accomplished by a dehydration to enzyme-bound *cis*-aconitate, which subsequently is hydrated to give isocitrate. Next, isocitrate undergoes oxidation

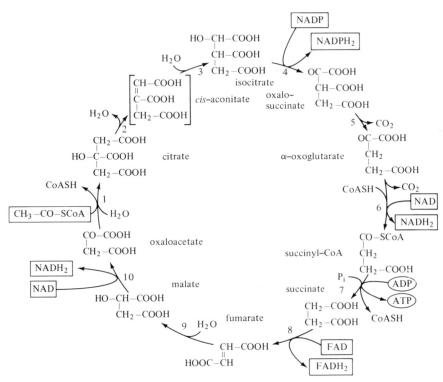

Figure 2.5. Oxidation of acetyl-CoA via the tricarboxylic acid cycle. 1, citrate synthase; 2 and 3, cis-aconitate hydratase; 4 and 5, isocitrate dehydrogenase; 6, α-oxoglutarate dehydrogenase complex; 7, succinate thiokinase; 8, succinate dehydrogenase; 9, fumarase; 10, malate dehydrogenase.

to oxalosuccinate, which is decarboxylated by the same enzyme, isocitrate dehydrogenase, to yield α-oxoglutarate. Like the corresponding enzyme of most other bacteria the isocitrate dehydrogenase of *E. coli* is NADP-specific. Oxidation of α-oxoglutarate to succinyl-CoA is catalyzed by an enzyme complex analogous to the one involved in the oxidation of pyruvate to acetyl-CoA. It also contains three enzyme species, one catalyzing the decarboxylation of TPP-bound α-oxoglutarate, one accepting the succinyl moiety and releasing succinyl-CoA, and one with dihydrolipoic dehydrogenase activity. The dihydrolipoic dehydrogenase is identical to the corresponding component of the pyruvate dehydrogenase complex, and it has been shown that *E. coli* contains a single gene locus for this enzyme. Thus, newly synthesized dihydrolipoic dehydrogenase will combine with two enzyme components to yield either the pyruvate or the α-oxoglutarate dehydrogenase complex.

In the next step the energy of the thioester bond of succinyl-CoA is used to synthesize ATP from ADP and inorganic phosphate. This is another reaction in which ATP is generated by substrate-level phosphorylation.

$$\text{succinyl-CoA} + P_i + \text{ADP} \xrightarrow[\text{thiokinase}]{\text{succinate}} \text{succinate} + \text{ATP} + \text{CoA}$$

The corresponding mammalian enzyme phosphorylates GDP and IDP but not ADP. Succinate is oxidized to fumarate by succinate dehydrogenase. The enzyme resides in particles associated with the cytoplasmic membrane and transfers electrons from succinate to bound FAD. As will be discussed later the electrons are then channeled from FAD into the respiratory chain. That NAD is not used as electron acceptor in this oxidation reaction is because the fumarate/succinate system ($E_0' = +0.03$ V) has a more positive oxidation-reduction potential as compared to that of NAD/NADH$_2$ ($E_0' = -0.32$ V). Thus, succinate is a very weak reductant for NAD. Protein-bound FAD with E_0' of approximately -0.06 V is much more suitable.

Two additional enzymes are necessary to generate oxaloacetate. First fumarase hydrates fumarate to L-malate, then L-malate dehydrogenase oxidizes L-malate to oxaloacetate with NAD as H-acceptor.

With the oxidation of two acetyl-CoA via the tricarboxylic acid cycle glucose is completely oxidized to CO_2. The hydrogen which theoretically is available in this oxidation is conserved in the form of reduced coenzymes. The number of **available hydrogens** is usually calculated by oxidizing the substrate (on paper) to CO_2 with water; for glucose it is 24(H):

$$\underset{\text{glucose}}{C_6H_{12}O_6} + 6H_2O \longrightarrow \underset{\text{available hydrogen}}{24(H)} + 6CO_2$$

$$\text{glucose} + 8\text{NAD} + 2\text{NADP} + 2\text{FAD} + 4\text{ADP} + 4P_i \longrightarrow$$
$$\underbrace{8\text{NADH}_2 + 2\text{NADPH}_2 + 2\text{FADH}_2}_{24(H)} + 4\text{ATP} + 6CO_2$$

It is clear that the process of glucose oxidation soon would come to a standstill if there were not reactions to regenerate the oxidized forms of the coenzymes. As in other aerobic organisms the principal H-acceptor during aerobic growth of E. coli is oxygen and the apparatus used to react the reduced forms of the coenzymes with oxygen is the respiratory chain. However, in this connection it must be mentioned that E. coli is a facultative anaerobe and that even under aerobic conditions a part of the glucose is catabolized via fermentative pathways not involving oxygen. For the sake of simplicity the simultaneously occurring fermentative metabolism of E. coli will be neglected here and will be discussed in a later chapter.

V.   The Formation of ATP in the Respiratory Chain

A.   Oxidation-reduction potential

An oxidation-reduction (OR) reaction may be written as follows:

$$A_{red} \rightleftharpoons A_{ox} + n \text{ electrons}$$
$$B_{ox} + n \text{ electrons} \rightleftharpoons B_{red}$$

If an equimolar solution of A_{red}/A_{ox} is added to an equimolar solution of B_{red}/B_{ox}, the direction of any reaction depends on the tendency of the A_{red}/A_{ox} system to donate electrons to the B_{red}/B_{ox} system and vice versa. A quantitative measure of this "tendency" in OR systems is their redox potential.

To measure redox potentials the hydrogen electrode—a solution containing H^+ of unit activity (pH=0) and an inert metal electrode in equilibrium with H_2 at 1 atm—is normally used as reference electrode. Its oxidation-reduction potential is arbitrarily taken to be zero ($E_0 = 0$ volt). Spontaneously or in the presence of the appropriate catalysts, OR systems with negative redox potentials reduce H^+ to hydrogen. OR systems with a positive E_0 are reduced by H_2. The dependency of the oxidation-reduction potential on the concentration of the components of the OR system is expressed by the Nernst equation:

$$E = E_0 + \frac{R \cdot T}{n \cdot F} \cdot \ln \frac{[\text{ox}]}{[\text{red}]}$$

(R, gas constant; T, absolute temperature; n, number of electrons; F, Faraday constant).

In all reactions involving protons the standard oxidation-reduction potential refers to pH=0. In that most biological reactions proceed at pH values near 7 it is more practical to calculate the standard oxidation-reduction potential of biological systems when the pH is 7.

At pH 7 and 30°C the potential of the hydrogen electrode becomes −0.42 volts:

$$\underset{(\text{pH}=7)}{E'_0} = \underset{(\text{pH}=0)}{E_0} + \frac{R \cdot T}{n \cdot F} \cdot \ln 10^{-7}$$

$$E'_0 = 0 + \frac{8.314 \times 303}{1 \times 96494} \times 2.303 \times -7 = -0.42 \text{ V}$$

The E'_0 values of NAD and O_2 are −0.32 and +0.81 V, respectively, and the difference between them is the potential span of the respiratory chain.

B.   Components of the respiratory chain

The major components of the respiratory chain are proteins bearing prosthetic groups with oxidation-reduction potentials lying between those of NAD and oxygen. In the mitochondrial membrane of eukaryotic organisms and the cytoplasmic membrane of bacteria, these proteins are arranged in such a way that the reducing power of $NADH_2$ can flow to oxygen via carriers of increasing oxidation-reduction potentials as if over cascades. However, the composition of the *E. coli* respiratory chain is not identical to that of mitochondria (Figure 2.6). As is indicated by the rectangles the mitochondrial chain contains four complexes, which can be isolated as

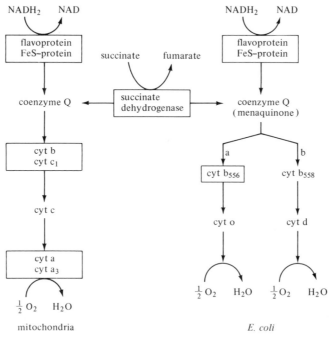

Figure 2.6. Components of the respiratory chains of mitochondria and of *E. coli*. The *E. coli* chain is branched; a dominates in cells growing under real aerobic conditions, b dominates in oxygen-limited cells. FeS-protein, iron-sulfur protein; cyt, cytochrome.

such. Complex 1 is the $NADH_2$ dehydrogenase; it contains flavin mononucleotide (FMN) and iron-sulfur proteins and transfers hydrogen from $NADH_2$ to coenzyme Q. Succinate dehydrogenase (complex 2) also feeds hydrogen into the respiratory chain at the coenzyme Q level. This enzyme is a flavin adenine dinucleotide (FAD)-containing protein. Cytochrome c is then reduced by coenzyme Q via complex 3 and finally, complex 4 catalyzes the transfer of reducing power to oxygen.

Many details about the structural organization of the carriers are not yet known for the *E. coli* respiratory chain. However, its composition is already known to be different from that of mitochondria; cytochrome c is not involved, and, most noteworthy, the *E. coli* chain is branched.

In cells growing under fully aerobic conditions reducing power flows preferentially via coenzyme Q, cytochrome b_{556} and cytochrome o to oxygen. Oxygen-limited cells employ coenzyme Q or menaquinone and

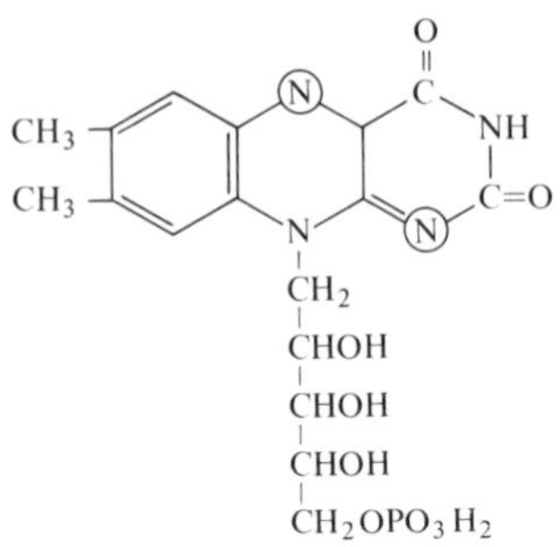

Figure 2.7. Flavin mononucleotide (FMN), the prosthetic group of the $NADH_2$ dehydrogenase of the respiratory chain. Circles indicate where reduction takes place. Many other enzymes including succinate dehydrogenase contain flavin adenine dinucleotide (FAD). The oxidation-reduction potential of flavoproteins is not identical with the potentials of FMN and FAD ($E_0' = -0.19$ and -0.22 V, respectively). Due to interaction of the protein with its prosthetic groups E_0' can be either more negative or more positive.

cytochromes b_{558} and d as carriers. It is assumed that the latter route yields less ATP by electron transport phosphorylation than route a.

   The structural formulas of the four types of carriers present in respiratory chains are given in Figures 2.7 to 2.10. Flavoproteins, coenzyme Q, and menaquinone are **hydrogen carriers**. During reduction two hydrogens are transferred to one carrier molecule:

$$\text{carrier} + 2H \rightleftharpoons \text{carrier} - H_2$$

   Cytochromes and iron-sulfur proteins are **electron carriers**. During reduction one electron is transferred to the central iron of the cytochrome

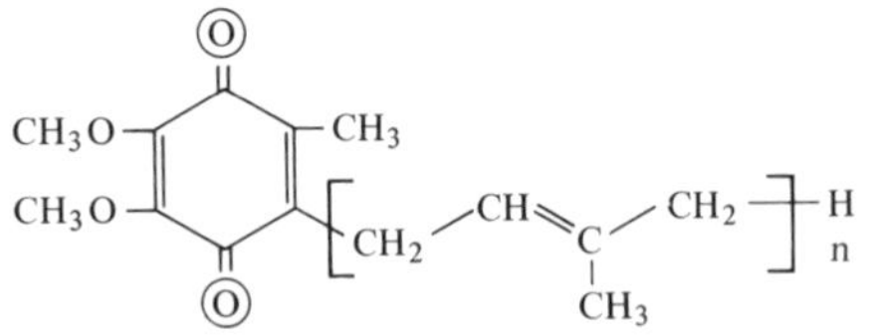

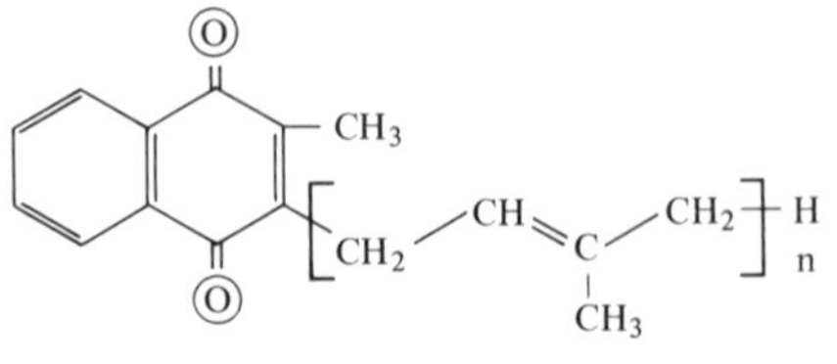

Figure 2.8. Coenzyme Q (a) (ubiquinone) and menaquinone (b). Circles indicate where reduction takes place. n varies from 4 to 10; in *E. coli*, $n=8$ for both quinones.

$$\begin{array}{ccc}
CH_3 & CH_3 & CH_3 \\
| & | & | \\
CH_2-CH-(CH_2)_3-CH-(CH_2)_3-CH \\
| & & | \\
HO-CH & & CH_3
\end{array}$$

CH₃

H₃C— —CH=CH₂

N⋯⋯N

Fe

N N

O=C—
|
H

CH₂   CH₂
| |
HOOC—CH₂   CH₂—COOH

Figure 2.9. Prosthetic group of a cytochrome. Cytochromes differ in the substituents of the two upper rings. Their E_0' values are different and depend on the electron affinity of protein ligands interacting with the central Fe atom.

and to one of the irons of the FeS cluster of the iron-sulfur protein, respectively:

$$\text{carrier} - Fe^{3+} + H \rightleftharpoons \text{carrier} - Fe^{2+} + H^+$$

Thus, if electron carriers are reduced by hydrogen carriers, protons are released. Conversely, the reduction of hydrogen carriers by electron carriers requires protons. This is important, because hydrogen and electron carriers arranged in alternating sequence in a membrane may cause **proton translocations**; protons released may be excreted at one side of the membrane and

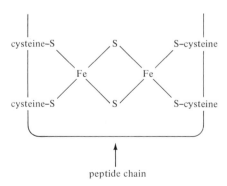

Figure 2.10. The FeS cluster of an iron-sulfur protein. Iron is bound to cysteine residues of the peptide chain and to sulfide.

protons required may be taken up from the other side:

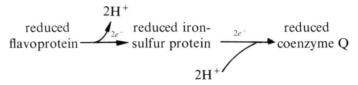

Proton translocations are the basis for the chemiosmotic hypothesis of ATP formation during respiration.

C. Electron transport phosphorylation

A 0.1-volt increase of the E_0' value during electron transport from one carrier to another corresponds to a free-energy change of $\Delta G_0' = -4.6$ kcal (-19.2 kJ) per mole. The value for the oxidation of $NADH_2$ with oxygen ($\Delta E_0' = 1.13V$) is then $\Delta G_0' = -52$ kcal (-217.6 kJ) per mole. If the respiratory chain contained only those components discussed thus far, this energy would have to be released as heat. Yet, it has been known for more than 30 years from the work with mitochondrial systems that part of this energy is used to synthesize ATP from ADP and inorganic phosphate. This process is called **oxidative** or **electron transport phosphorylation** and was first demonstrated by Kalckar and Belitser with minced liver and muscle tissue. Electron transport phosphorylation studies *in vitro* are very difficult, as during the preparation of subcellular fractions soluble components are easily separated from the matrix-bound respiratory chain. Thus sophisticated methods are necessary to reconstitute a chain able to catalyze simultaneously the oxidation of $NADH_2$ and an efficient electron transport phosphorylation. Through the work of Green, Racker, Lehninger, Chance, and others it is now known that there are three phosphorylation sites at the respiratory chain of mitochondria. Thus, three ADP can be phosphorylated per two electrons transferred from $NADH_2$ to oxygen. The number of ADPs phosphorylated per atom oxygen is frequently expressed as the **P:O ratio**; in the mitochondrial chain it is 3 for $NADH_2$ as electron donor and 2 for $FADH_2$ as electron donor. Even the sites of ATP formation are known: the dehydrogenation of $NADH_2$, the oxidation of cytochrome b, and the oxidation of cytochrome a (see Figure 2.6).

The bacterial respiratory chain seems to be even more sensitive to isolation procedures, and subcellular fractions often have almost completely lost the ability to couple electron transport with the phosphorylation of ADP. The first P:O ratios reported for bacterial systems were generally below 1 and it was assumed that electron transport phosphorylation in bacteria is less efficient than in mitochondria. With improved methods, which largely prevented structural damage of the particles and which included the readdition of soluble factors, it has been possible to obtain P:O ratios greater than 2 for a number of bacterial species. Thus it appears safe to conclude that at

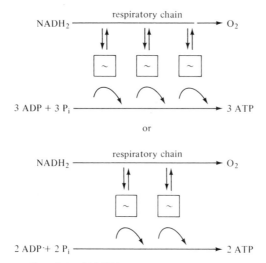

Figure 2.11. Electron flow from $NADH_2$ to oxygen generates an energized state of the components of the respiratory chain.

least some aerobic bacteria are able to oxidize $NADH_2$ via the respiratory chain with a P:O ratio of 3; for *E. coli* this value is probably 2, and the sites of ATP formation are the dehydrogenation of $NADH_2$ and the oxidation of one of the cytochromes.

How then is ATP synthesized? Although great efforts have been made in many laboratories over the years, the exact mechanism of electron transport phosphorylation is as yet not fully understood. Hydrogen and electron flow through the respiratory chain must generate an energized state ("~") of two or three components within the chain, which may provide the energy for ATP synthesis when returning to the normal state (Figure 2.11).

The question thus seems to be, what is "~"? Three hypotheses have been elaborated for the mitochondrial respiratory chain.

1. Chemical coupling hypothesis. This hypothesis proposed by Lehninger uses well-known enzymatic reactions as models. As is indicated in Figure 2.12, the oxidation of a carrier of the respiratory chain (A) is coupled to the formation of a linkage (A~J) with a high free energy of hydrolysis. Reactions of this type are known from the glyceraldehyde-3-phosphate dehydrogenase and pyruvate dehydrogenase reactions. By subsequent transferase reactions the carrier A is regenerated and a phosphorylated enzyme (E~P) is formed, which finally phosphorylates ADP to ATP.

2. Mechanochemical or conformational coupling hypothesis. The transformation of chemical work (ATP hydrolysis) into mechanical work is well known from studies of muscle contraction, and a reverse process might possibly be involved in electron transport phosphorylation: electron flow through the respiratory chain could lead to conformational changes in

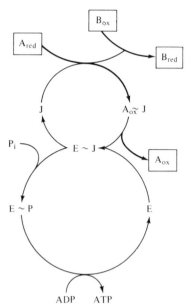

Figure 2.12. Chemical coupling hypothesis. During transfer of electrons from carrier A to carrier B compound J reacts with A to form the high-energy compound $A \sim J$. A transferase reaction between E and $A \sim J$ yields A and $E \sim J$. Then J is displaced by inorganic phosphate. Finally, $E \sim P$ phosphorylates ADP to yield ATP.

certain components of the chain, and their return from the strained energized state to the normal one might result in the concomitant formation of ATP from ADP and P_i. Volume changes of mitochondria can indeed be observed.

3. Chemiosmotic hypothesis. In 1961 Mitchell proposed this very attractive hypothesis. It assumes that (1) the membrane which harbors the respiratory chain is impermeable to OH^- and H^+, (2) the respiratory chain is localized in the membrane in such a way that a pH gradient and a membrane potential are formed by vectorial extraction and excretion of protons during electron transport, and (3) the ATPase is so ingeniously constructed that it can take advantage of the **protonmotive force** (sum of chemical gradient of protons and membrane potential). Mitchell's hypothesis is illustrated in Figure 2.13. It has been already mentioned that alternating sequences of hydrogen and electron carriers may cause proton translocations. So it is assumed that the carriers are arranged in the membrane in the form of loops with an electron carrier facing the outer part and a hydrogen carrier facing the inner part of the membrane [Figure 2.13(a)]. A pH gradient and a membrane potential are established during oxidation of $NADH_2$ [Figure 2.13(b)]. It allows ATP synthesis from ADP and inorganic phosphate at an enzyme (BF_0F_1), which is oriented in the membrane in such a way that the protons formed are trapped by OH^- groups inside and the hydroxyl ions formed are released

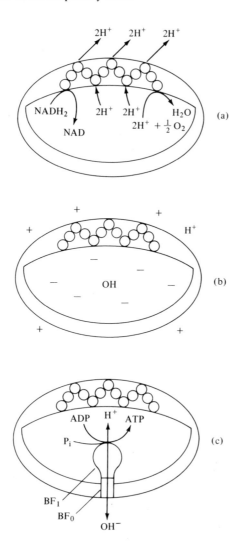

Figure 2.13. Chemiosmotic hypothesis for oxidative phosphorylation in mitochondria. [F. M. Harold, *Bacteriol. Rev.* **36**, 172–230 (1972).] The oxidation of $NADH_2$ via the respiratory chain leads to the extraction of protons from the intramitochondrial compartment (a). As the consequence a pH gradient and a membrane potential are established (b). The removal of H^+ and OH^- from ADP and inorganic phosphate to yield ATP is then accomplished by the enzyme BF_0F_1. This enzyme is oriented in the membrane in such a way that the protons are trapped by OH^- groups inside and the hydroxyl ions by H^+ outside (c). BF_1 exhibits Mg-ATPase activity.

outside [Figure 2.13(c)]:

$$ADP + P_i \xrightarrow{\ \ BF_0F_1\ \ } ATP + H^+\!\uparrow + OH^-\!\downarrow$$

Enzyme BF_0F_1 consists of two components, BF_0 which is embedded in the membrane and BF_1 which is attached to BF_0. The component BF_1 is also called coupling factor. It can be released from the membrane, and it exhibits magnesium ion-dependent ATPase activity.

A scheme of the functional organization of the redox carriers in the respiratory chain of *E. coli*, as has been proposed by Haddock and Jones, is shown in Figure 2.14. It indicates that this chain comprises only two proton extraction sites and not three as in the case of the mitochondrial chain. Therefore, the P:O ratio of the *E. coli* chain can be only two.

From our knowledge of mechanisms of enzyme-catalyzed reactions, chemical coupling seems easy to accept. However, this hypothesis does not assign any functional importance to the membrane. Furthermore, the postulated high-energy intermediates $(A \sim J; E \sim J)$ have never been detected. Conformational coupling is somewhat related to chemical coupling, but it replaces a defined intermediate $(A \sim J)$ with an energy-rich conformational state. The chemiosmotic hypothesis ascribes an important role to the

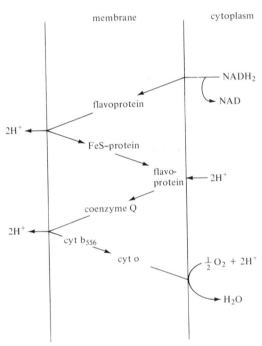

Figure 2.14. Functional organization of the components of the *E. coli* respiratory chain. [B. A. Haddock and C. W. Jones, *Bacteriol. Rev.* 41, 47–99 (1977)]. Not shown in this scheme is the less ATP-yielding branch with the cytochromes b_{558} and d (see Figure 2.6).

membrane (pH gradient, membrane potential) and assumes the presence of a membrane-bound ATPase, which indeed is present in mitochondria and bacteria.

It should be pointed out that most of the experiments concerning the mechanism of electron transport phosphorylation have been done with mitochondrial systems. However, because of many similarities between the mitochondrial and bacterial chain it seems probable that the mechanism of electron transport phosphorylation in bacteria and mitochondria is basically the same. Furthermore, it has been suggested that the mitochondria of eukaryotic organisms have evolved from microbial symbionts. If true this would also imply similarities in the mechanisms of electron transport phosphorylation in mitochondria and bacteria.

D.   Uncouplers and inhibitors

In growing cells of *E. coli* the oxidation of $NADH_2$ and the phosphorylation of ADP are coupled. This coupling is largely lost when subcellular fractions are prepared. As a result, the $P:O$ ratio is low and only little ATP is formed during oxidation of $NADH_2$. *In vivo* this uncoupling can also be achieved by the addition of certain compounds to cell suspensions. The classical uncoupling agent introduced by Loomis and Lipman is **2,4-dinitrophenol**. Glucose is oxidized almost completely in the presence of 2,4-dinitrophenol, but ATP is not formed and cell material cannot be synthesized. In the absence of 2,4-dinitrophenol about 50% of the carbon of glucose is released as CO_2 with the remaining 50% being converted into cellular material:

$$C_6H_{12}O_6 + 3O_2 \xrightarrow{\text{growth}} \text{cell material} + 3CO_2 + 3H_2O$$

$$C_6H_{12}O_6 + 6O_2 \xrightarrow{\text{dinitrophenol}} 6CO_2 + 6H_2O$$

Certain dyes such as **methylene blue** can replace oxygen as electron acceptor and allow the oxidation of organic compounds to CO_2 in the absence of oxygen. Methylene blue accepts hydrogen from flavoproteins and is reduced to its leuco form:

methylene blue

leuco form

The time required to reduce a fixed amount of methylene blue has been used as a measure of the capacity of washed *E. coli* suspensions to oxidize glucose and other organic substrates (Thunberg technique).

The antibiotic **oligomycin** inhibits electron transport phosphorylation and thus blocks respiration when it is coupled to the phosphorylation of ADP (in the presence of an uncoupler, oligomycin does not inhibit respiration).

Compounds like azide, cyanide, CO, and certain barbiturates inhibit hydrogen or electron transfer at one or more sites of the respiratory chain. All these agents have been valuable tools for the elucidation of respiration and electron transport phosphorylation.

E.   The importance of superoxide dismutase

Concentrations of oxygen higher than that found in air are toxic to many aerobic microorganisms. Obligate anaerobes such as clostridia and methanogenic bacteria die when they are exposed to air. For some time this deleterious effect of oxygen was related exclusively to the accumulation of its reduction product hydrogen superoxide (H_2O_2). The latter is formed whenever activated hydrogen in the form of reduced flavoproteins or reduced iron-sulfur proteins come together with oxygen and oxidases that are present in all organisms.

$$FADH_2 + O_2 \xrightarrow{\text{oxidase}} FAD + H_2O_2$$

Aerobes, therefore, contain catalase, which converts H_2O_2 to oxygen and water:

$$2H_2O_2 \xrightarrow{\text{catalase}} 2H_2O + O_2$$

Catalase is indeed very important. However, it is now apparent that a more toxic compound than H_2O_2 is produced from oxygen in biological systems. This compound is the superoxide radical (O_2^-), which is formed by a univalent reduction of oxygen with reduced flavins, quinones, or other electron carriers. *E. coli* and all aerobic and aerotolerant microorganisms contain the enzyme superoxide dismutase, which converts the radical to H_2O_2 and O_2.

$$O_2^- + O_2^- + 2H^+ \xrightarrow{\substack{\text{superoxide} \\ \text{dismutase}}} H_2O_2 + O_2$$

Superoxide dismutase of *E. coli* is a red manganoprotein whereas the enzyme of mammals contains copper or zinc.

VI. Summary

1. *Escherichia coli* contains a PEP phosphotransferase system, which is responsible for the uptake of glucose. Transport of the sugar into the cell is coupled to its phosphorylation to glucose-6-phosphate.

2. Glucose-6-phosphate is degraded to pyruvate via the Embden–Meyerhof–Parnas pathway. Key intermediate of this pathway is fructose-1,6-bisphosphate. In the conversion of glucose-6-phosphate to pyruvate ATP is required for the phosphofructokinase reaction and ATP is produced in the 3-phosphoglycerate and pyruvate kinase reactions.

3. The oxidative decarboxylation of pyruvate to acetyl-CoA is accomplished by the pyruvate dehydrogenase complex. This complex consists of three enzymes: a pyruvate dehydrogenase, a dihydrolipoic transacetylase, which transfers the acetyl residue to coenzyme A, and a dihydrolipoic dehydrogenase, which transfers hydrogen from enzyme 2 to NAD.

4. Acetyl-CoA is oxidized in the tricarboxylic acid cycle. The oxidation yields $NADPH_2$ in the isocitrate dehydrogenase reaction, $NADH_2$ in the α-oxoglutarate and malate dehydrogenase reactions, and $FADH_2$ in the succinate dehydrogenase reaction. ATP is formed in the conversion of succinyl-CoA to succinate.

5. The respiratory chains of *E. coli* and of mitochondria are different. The *E. coli* chain does not contain a cytochrome of the c-type; furthermore, it is branched. The transport of two electrons from $NADH_2$ to oxygen is coupled to the phosphorylation of probably two ADPs. For the mitochondrial chain and the chain of some other bacteria this value is three.

6. The mechanism underlying the formation of ATP by electron transport phosphorylation is not completely established. Many experimental results favor the chemiosmotic hypothesis of Mitchell. Electron transport establishes a protonmotive force at a membrane. This force provides the energy for ATP synthesis at the enzyme BF_0F_1.

7. During biological oxidations the toxic superoxide radical O_2^- is formed in small concentrations. Superoxide dismutase converts this compound into H_2O_2 and O_2.

8. The overall equation for glucose oxidation by *E. coli* is:

$$glucose + 8NAD + 2NADP + 2FAD + 4ADP + 4P_i \longrightarrow$$
$$8NADH_2 + 2NADPH_2 + 2FADH_2 + 4ATP + 6CO_2$$

The oxidation of $10NAD(P)H_2$ via the respiratory chain yields maximally 20ATP and the oxidation of $2FADH_2$ yields maximally 2ATP.

Chapter 3

Biosynthesis of *Escherichia coli* Cells from Glucose

It has already been mentioned that *E. coli*—when growing aerobically on glucose—oxidizes about 50% of the glucose to CO_2 for the production of ATP. The remaining 50% is converted into cellular material. It is this conversion that consumes most of the ATP formed in oxidation. In this chapter the principal biosynthetic reactions involved in synthesis of cellular material and the main sites of ATP consumption will be outlined.

I. Composition of *E.coli* Cells

More than 95% of the cellular material of *E. coli* and of other microorganisms consists of macromolecules. A typical analysis of microbial cells is given in Table 3.1. Proteins account for approximately 52% and nucleic acids for 19% of the dry weight. About 3% of the dry weight of the cells includes low-molecular-weight organic compounds and salts.

Figure 3.1 presents a general scheme for the formation of the cellular constituents. Intermediates of glucose breakdown (hexose phosphates, PEP, pyruvate, acetyl-CoA, oxaloacetate, α-oxoglutarate) are used as starting material to make all the required amino acids, vitamins, sugar phosphates, fatty acids, ribo-, and deoxyribonucleotides. Polymerization reactions then lead to the formation of macromolecules.

Most of the biosynthetic reactions of *E. coli* are now known, and it is thus possible to calculate the amount of ATP required for the formation of the macromolecules (Table 3.2). At first glance it is surprising to learn that from 34.8 mmol of ATP necessary for the synthesis of 1 g of cell material, 19.1 mmol (56%) have to be invested in polymerization of amino acids. A comparatively small percentage of the available ATP is required for the formation of the amino acids from glucose. During RNA and DNA synthesis, the

Table 3.1. Content of macromolecules of microbial cells[a]

macromolecule	amount (g/100 g of dried cells)
protein	52.4
polysaccharide	16.6
lipid	9.4
RNA	15.7
DNA	3.2
total	97.3

[a] A. H. Stouthamer, *Antonie van Leeuwenhoek* **39**, 545–565 (1973).
Note: These figures of composition are representative but the actual composition changes markedly among different (e.g., Gram-positive versus Gram-negative) bacteria and varies with differing growth conditions (e.g., RNA content).

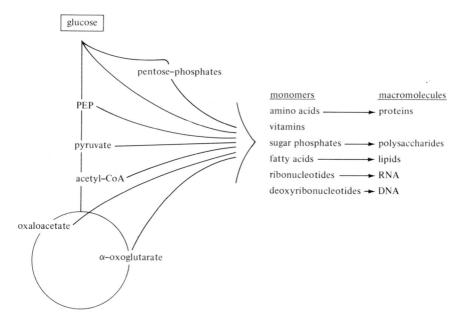

Figure 3.1. General scheme of the biosynthesis of cell material from glucose.

Table 3.2. ATP requirement for the formation of microbial cells from glucose and inorganic salts[a]

macromolecule	ATP (mmol) required for the synthesis of the macromolecule content of 1 g of dried cells
polysaccharide	2.1
protein	
glucose→amino acids	1.4
polymerization	19.1
lipid	0.1
RNA	
glucose→nucleoside monophosphates	3.5
polymerization	0.9
DNA	
glucose→deoxynucleoside monophosphates	0.9
polymerization	0.2
ATP required for transport processes (salts)	5.2
ATP required for RNA turnover	1.4
total ATP requirement	34.8

[a]A. H. Stouthamer, *Antonie van Leeuwenhoek* **39**, 545–565 (1973).

reactions leading from glucose to the appropriate nucleoside monophosphates are more ATP-consuming than the final polymerization reactions. A considerable amount of ATP is also required to compensate for RNA turnover. The half-life of messenger RNA is approximately 3 min and during the time the cell mass of a culture doubles, the mRNA content of the cells has to be synthesized several times. Finally, about one-seventh of the ATP is invested in transport processes; of this the greatest proportion is required for the uptake of ammonium ions.

In conclusion, a culture of *E. coli* growing aerobically on glucose has to invest the major amount of ATP gained by respiration in the polymerization of amino acids. Smaller amounts of ATP flow into the biosynthesis of monomers, other polymerization reactions, and transport processes. Not considered here are the energy of maintenance, which depends very much on the growth conditions and may represent 10 to 20% of the ATP-consuming reactions in growing cells and the energy of flagellar movement, which is comparatively small.

In the following sections some of the biosynthetic reactions of *E. coli* will be discussed.

II.   Assimilation of Ammonia

If the concentration of ammonia in the environment of the *E. coli* cells is high, it is assimilated by reductive amination of an intermediate of the tricarboxylic acid cycle, α-oxoglutarate. The enzyme catalyzing this reaction is a NADP-specific **L-glutamate dehydrogenase.** From the L-glutamate thus formed the amino group can then be transferred to α-oxoacids. *E. coli* contains two enzymes, **transaminase A and B**, responsible for catalysis of this amino group transfer; their low specificity allows the formation of more than 10 amino acids from the corresponding α-oxoacids.

Figure 3.2(a) shows the glutamate dehydrogenase reaction and the formation of valine from α-oxoisovalerate as an example of a transaminase reaction.

A second important reaction for the assimilation of ammonia is the L-glutamine synthetase reaction:

$$
\begin{array}{lcl}
\begin{array}{l}
\text{COOH} \\
| \\
\text{H}_2\text{N}-\text{CH} \\
| \\
\text{CH}_2 \;+\; \text{ATP} \;+\; \boxed{\text{NH}_3} \\
| \\
\text{CH}_2 \\
| \\
\text{COOH}
\end{array}
&
\xrightarrow[\text{L-glutamine synthetase}]{}
&
\begin{array}{l}
\text{COOH} \\
| \\
\text{H}_2\text{N}-\text{CH} \\
| \\
\text{CH}_2 \;+\; \text{ADP} \;+\; \text{P}_i \\
| \\
\text{CH}_2 \\
| \\
\text{CO}-\boxed{\text{NH}_2}
\end{array}
\end{array}
$$

Glutamine serves as NH_2 donor in the biosynthesis of a number of compounds such as purines, tryptophan, histidine, glucosamine-6-phosphate, and also in the formation of carbamyl phosphate, which is one of the precursors of the pyrimidine compounds:

$$
\begin{array}{lcl}
\begin{array}{l}
\text{COOH} \\
| \\
\text{H}_2\text{N}-\text{CH} \\
| \\
\text{CH}_2 \quad + \\
| \\
\text{CH}_2 \qquad 2\text{ATP} \\
| \\
\text{CO}-\boxed{\text{NH}_2}
\end{array}
&
\begin{array}{l}
\text{HCO}_3^- \\[1em]
\xrightarrow[\text{synthetase}]{\text{carbamyl phosphate}}
\end{array}
&
\begin{array}{l}
\text{COOH} \\
| \\
\text{H}_2\text{N}-\text{CH} \qquad \boxed{\text{H}_2\text{N}}-\text{CO}-\text{OPO}_3\text{H}_2 \\
| \\
\text{CH}_2 \quad + \\
| \\
\text{CH}_2 \qquad\qquad + 2\text{ADP} + \text{P}_i \\
| \\
\text{COOH}
\end{array}
\end{array}
$$

In recent years it has become clear that the glutamate dehydrogenase is not involved in the primary assimilation of ammonia at low concentrations of ammonia (<1 mM). Its K_m value for ammonia is high and the enzyme cannot function efficiently under these conditions. It has been shown, first by Tempest and later by other investigators, that at low NH_3 concentrations a combination of **glutamine synthetase** and **glutamate synthase** is responsible for glutamate formation from α-oxoglutarate and ammonia. The latter enzyme—which is frequently called **GOGAT** (glutamine: α-oxoglutarate

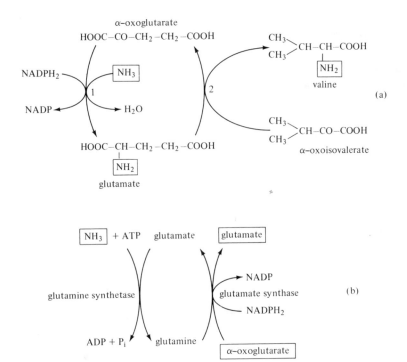

Figure 3.2. Assimilation of ammonia. (a) By glutamate dehydrogenase (1) and subsequent transfer of the amino group by a transaminase (2). (b) By glutamine synthetase/glutamate synthase (GS/GOGAT).

aminotransferase)—catalyzes the reductive transfer of the amido group of glutamine to α-oxoglutarate. The pathway, which is widespread among bacteria, is illustrated in Figure 3.2(b).

III.   Assimilatory Reduction of Sulfate

As in most bacteria *E. coli* uses sulfate as the principal sulfur source. However, this sulfate must be reduced since sulfur of the majority of sulfur-containing cellular compounds is at the oxidation level of H_2S. The path of sulfate reduction is summarized in Figure 3.3. First, sulfate is actively transported into the cell and then activated by the ATP sulfurylase reaction. Of the **adenosine-5′-phosphosulfate (APS)** and pyrophosphate thus formed, the latter may be hydrolyzed by a pyrophosphatase, thereby favoring APS synthesis. The subsequent phosphorylation of APS in the 3′-position yields **adenosine-3′-phosphate-5′-phosphosulfate (PAPS)**.

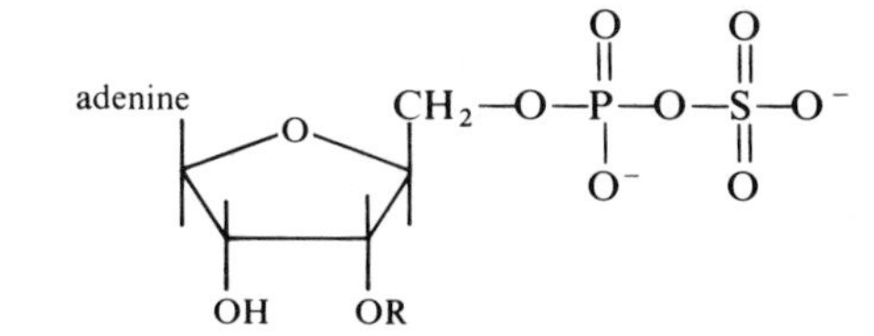

R = H, adenosine-5-phosphosulfate (APS)

R = PO_3H_2, adenosine-3′-phosphate-5′-phosphosulfate (PAPS)

The reduction of PAPS requires involvement of a thiol compound (thioredoxin) and leads to the release of sulfite. The enzyme sulfite reductase then catalyzes a 6-electron transfer from $3NADPH_2$ to sulfite with the formation of H_2S. Since this is very toxic to many microorganisms, the H_2S is immediately incorporated into O-acetylserine. The product formed, L-cysteine, is the most important precursor of sulfur-containing compounds of the cell.

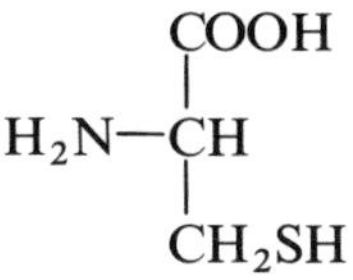

L-cysteine

The pathway just outlined is referred to as assimilatory sulfate reduction. A mechanistically different dissimilatory type of sulfate reduction is found in certain anaerobic bacteria (see Chapter 8).

$$\text{sulfate (outside)} \xrightarrow{\text{active transport}} \text{sulfate (inside)}$$

$$\text{ATP + sulfate} \xrightarrow{\text{ATP sulfurylase}} \text{adenosine-5'-phosphosulfate (APS)} + PP_i$$

$$\text{ATP + APS} \xrightarrow{\text{APS phosphokinase}} \text{adenosine-3'-phosphate-5'-phosphosulfate (PAPS) + ADP}$$

$$\text{2 RSH + PAPS} \xrightarrow{\text{PAPS reductase}} \text{sulfite + AMP-3'-phosphate + RSSR}$$

$$\text{sulfite} + 3\ NADPH_2 \xrightarrow{\text{sulfite reductase}} H_2S + 3\ NADP$$

$$\text{O-acetyl-L-serine} + H_2S \xrightarrow{\text{O-acetylserine sulfhydrylase}} \text{L-cysteine + acetate} + H_2O$$

Figure 3.3. Assimilatory reduction of sulfate and formation of cysteine. RSH stands for thioredoxin; its reduced form is regenerated from the oxidized form with $NADPH_2$.

IV.   Biosynthesis of Amino Acids

Twenty amino acids are required for the biosynthesis of proteins, and they are formed from the metabolic precursors listed in Table 3.3. It is evident from this table that only a few compounds serve as substrates in amino acid synthesis. Oxaloacetate, for instance, is the starting point for the synthesis of six amino acids, α-oxoglutarate is the precursor of four, and pyruvate of three amino acids.

An inspection of the metabolic routes leading to the individual amino acids reveals that common pathways are frequently employed. Figure 3.4 summarizes the reactions leading to the **oxaloacetate and pyruvate family** of amino acids. By simple transamination reactions aspartate is formed from oxaloacetate and alanine from pyruvate. Aspartate is converted into asparagine by asparagine synthetase. Aspartate semialdehyde is a central intermediate for the synthesis of lysine, threonine, and methionine, and it is formed from aspartate via aspartate-4-phosphate. Its condensation with pyruvate yields dihydrodipicolinate, which is the precursor of diaminopimelate and lysine. The reduction of aspartate semialdehyde gives homoserine, which can be further metabolized to methionine and threonine. The latter amino acid is not used only in protein synthesis; part of it is converted by

Table 3.3. Precursors for amino acid biosynthesis

precursor	amino acid
pyruvate	alanine valine leucine
oxaloacetate	aspartate asparagine methionine lysine threonine isoleucine
α-oxoglutarate	glutamate glutamine arginine proline
3-phosphoglycerate	serine glycerine cysteine
phospho-enolpyruvate erythrose-4-phosphate	phenylalanine tyrosine tryptophan
5-phosphoribosyl-1-pyrophosphate + ATP	histidine

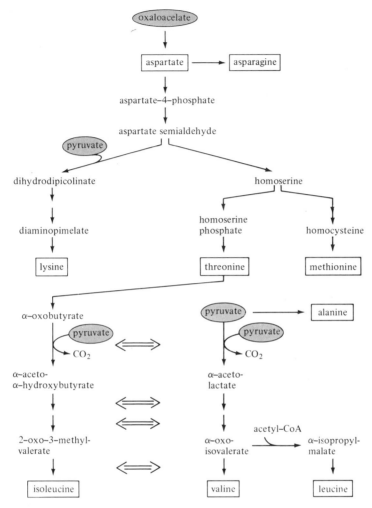

Figure 3.4. Biosynthesis of amino acids derived from oxaloacetate and pyruvate.

the enzyme threonine deaminase into α-oxobutyrate, which by a series of reactions, is converted into isoleucine. Interestingly enough the same set of enzymes, which catalyzes the reactions from α-oxobutyrate to isoleucine, also converts pyruvate into valine.

The biosynthesis of leucine branches off from the valine pathway at α-oxoisovalerate. Consequently, one set of enzymes is involved in the biosynthesis of three amino acids—isoleucine, valine, and leucine.

Figure 3.5 summarizes the reactions leading from 3-phosphoglycerate to **serine**, **glycine**, **and cysteine**. Dehydrogenation of 3-phosphoglycerate and subsequent action of a transaminase and a phosphatase yields L-serine. This amino acid can be converted into cysteine, as mentioned in connection

$$CH_3-CH_2-CO-COOH \quad ----\rightarrow$$

α-oxobutyrate

$$\begin{array}{c} H_3C-CH_2 \\ \diagdown \\ CH-CH-COOH \\ \diagup \quad | \\ H_3C \quad NH_2 \end{array}$$

isoleucine

⇑

4 enzymes

⇓

$$CH_3-CO-COOH \quad ----\rightarrow$$

pyruvate

$$\begin{array}{c} CH_3 \\ \diagdown \\ CH-CH-COOH \\ \diagup \quad | \\ CH_3 \quad NH_2 \end{array}$$

valine

with the assimilation of sulfate, or into glycine by the enzyme serine hydroxy-methyltransferase. In the course of this reaction carbon 3 of serine is transferred to tetrahydrofolic acid.

  Methylene tetrahydrofolic acid can be reduced to N^5-methyltetrahydro-folic acid, which functions as donor of methyl groups in a number of biosynthetic reactions, for instance in the synthesis of methionine from homocysteine.

  α-Oxoglutarate is the precursor of **glutamate** and **glutamine**, compounds

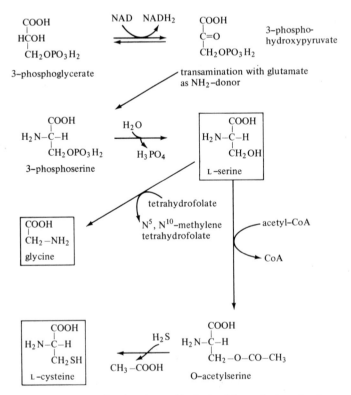

Figure 3.5. Biosynthesis of amino acids derived from 3-phosphoglycerate.

OH (5)
NH
(10)
CH—CH₂—NH—⟨⟩—CO—NH—CH
CH₂
H₂N N NH
tetrahydrofolic acid

COOH
CH₂
CH₂
COOH

serine
serine hydroxymethyltransferase
glycine

OH
CH₂—N—⟨⟩—CO—NH—CH
CH—CH₂
CH₂
H₂N N NH

COOH
CH₂
CH₂
COOH

N^5, N^{10}-methylene tetrahydrofolic acid

that serve not only as building blocks in protein synthesis, but also as important intermediates in nitrogen metabolism, as already outlined. In addition, glutamate is the precursor of **proline** and **arginine** (Figure 3.6).

Biosynthesis of the **aromatic amino acids** is very complicated; however, it has been of special interest to investigators since it involves the biosynthesis of the aromatic nucleus from aliphatic precursors. As illustrated in Figure 3.7 erythrose-4-phosphate and PEP are condensed to yield a C_7-compound,

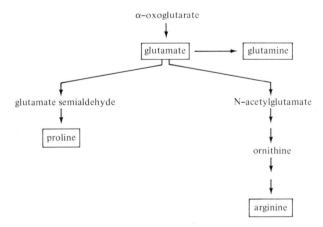

Figure 3.6. Biosynthesis of amino acids derived from α-oxoglutarate.

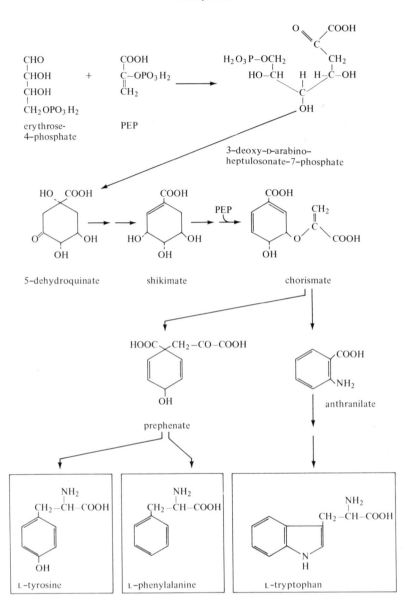

Figure 3.7. Biosynthesis of aromatic amino acids.

which undergoes cyclization (enzyme-catalyzed) to 5-dehydroquinate. The latter is converted into **chorismate**, a compound identified in 1964 as a common intermediate in the biosynthesis of aromatic amino acids. At this point the pathway branches off into two, one leading to tryptophan via anthranilate, and a second yielding prephenate, which is the precursor of both tyrosine and phenylalanine. When examining the scheme of aromatic

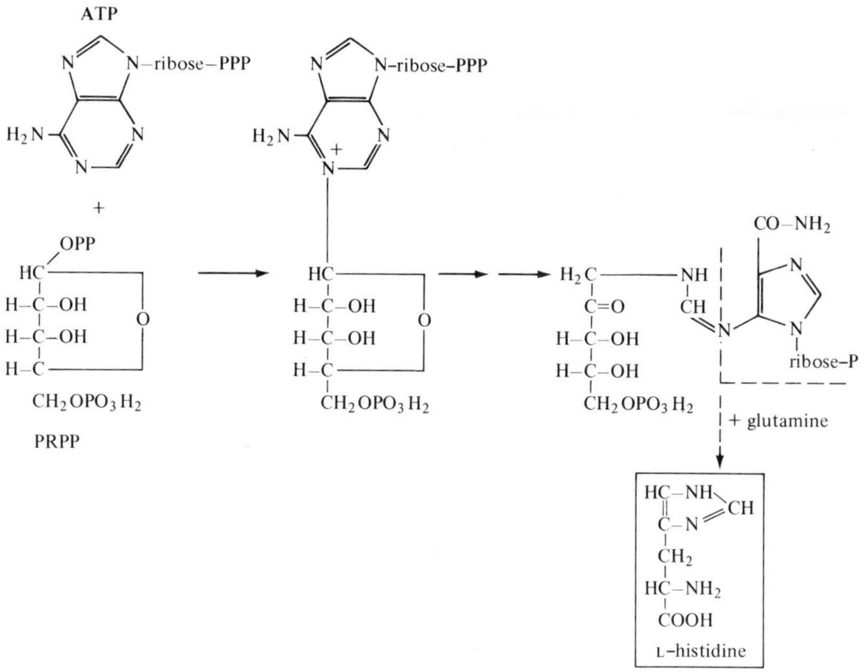

Figure 3.8. Origin of the histidine skeleton from 5-phosphoribosyl-1-pyrophosphate, ATP, and glutamine.

amino acid synthesis it is possible to understand how Davis was able to isolate a mutant of *E. coli* requiring phenylalanine, tyrosine, and tryptophan (*p*-aminobenzoate in addition) and why this mutant could grow in the presence of shikimate. This compound is an intermediate in the formation of all three amino acids; thus the mutant was obviously defective in one of the enzymes involved in the formation of shikimate from erythrose-4-phosphate and PEP.

A pathway completely independent of the reactions discussed thus far is employed for **histidine** biosynthesis. It is composed of nine enzymes, which manage the formation of histidine from 5-phosphoribosyl-1-pyrophosphate (PRPP), ATP, and glutamine. The resulting amino acid skeleton contains the five carbons of 5-phosphoribosyl-1-pyrophosphate, one carbon and one nitrogen of the pyrimidine ring of ATP, and the amido-nitrogen of glutamine (Figure 3.8).

V.   How Pentose Phosphates and $NADPH_2$ Are Formed

Many hydrogenation reactions in biosynthetic pathways require $NADPH_2$ as reducing agent and not $NADH_2$; the latter, however, is formed in most of the redox reactions involved in the oxidation of glucose to CO_2. Only the isocitrate dehydrogenase reaction yields $NADPH_2$. As in animals and a number of other microorganisms, *E. coli* regenerates additional $NADPH_2$ by the **oxidative pentose phosphate cycle**. This same pathway also serves as the source of pentose phosphates for the cell, one of which (5-phosphoribosyl-1-pyrophosphate) is required for the biosynthesis of nucleotides.

In the oxidative pentose phosphate cycle three series of reactions can be distinguished:

1. Oxidation of glucose-6-phosphate to ribulose-5-phosphate and formation of $NADPH_2$ and CO_2. This is accomplished by the enzymes glucose-6-phosphate dehydrogenase, a lactonase, and 6-phosphogluconate dehydrogenase.
2. Enzyme reactions which allow the formation of ribose-5-phosphate and xylulose-5-phosphate from ribulose-5-phosphate. The enzymes involved are phosphoribose isomerase and ribulose-5-phosphate-3-epimerase.
3. Transaldolase and transketolase reactions, which catalyze the formation of hexose-6-phosphates from pentose-5-phosphates.

Figure 3.9 summarizes all of the reactions of the oxidative pentose phosphate cycle. First glucose-6-phosphate is oxidized to 6-phosphogluconate by a NADP-specific glucose-6-phosphate dehydrogenase. Then the action of 6-phosphogluconate dehydrogenase leads to the reduction of a second NADP and to the formation of ribulose-5-phosphate and CO_2 with 3-keto-6-phosphogluconate as an intermediate in the reaction. Ribulose-5-phosphate can be isomerized to yield ribose-5-phosphate and epimerized to yield xylulose-5-phosphate. If *E. coli* were to need $NADPH_2$ and pentose phosphates in a ratio of 2:1 additional reactions would not be required. The sum would, therefore, be:

$$3 \text{ glucose-6-phosphate} + 6NADP \longrightarrow$$
$$3 \text{ pentose-5-phosphate} + 3CO_2 + 6NADPH_2$$

Now, consider a situation where there is a demand for more than $2NADPH_2$ for each pentose phosphate made; the excess pentose phosphate would have to be removed. This is accomplished by the enzymes **transketolase** and **transaldolase**, which convert pentose phosphates back into hexose phosphates. Both enzymes contain tightly bound thiamin pyrophosphate as coenzyme. Transketolase catalyzes the transfer of a glycolaldehyde group from xylulose-5-phosphate to ribose-5-phosphate or erythrose-4-phosphate:

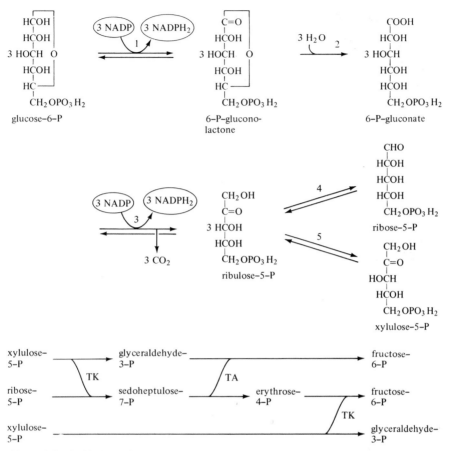

Figure 3.9. Oxidation of glucose-6-phosphate to ribulose-5-phosphate and conversion of pentose phosphate into hexose phosphate. 1, glucose-6-phosphate dehydrogenase; 2, lactonase; 3, 6-phosphogluconate dehydrogenase; 4, phosphoribose isomerase; 5, ribulose-5-phosphate-3-epimerase; TK, transketolase; TA, transaldolase.

Transaldolase catalyzes the transfer of a dihydroxyacetone group from sedoheptulose-7-phosphate to glyceraldehyde-3-phosphate:

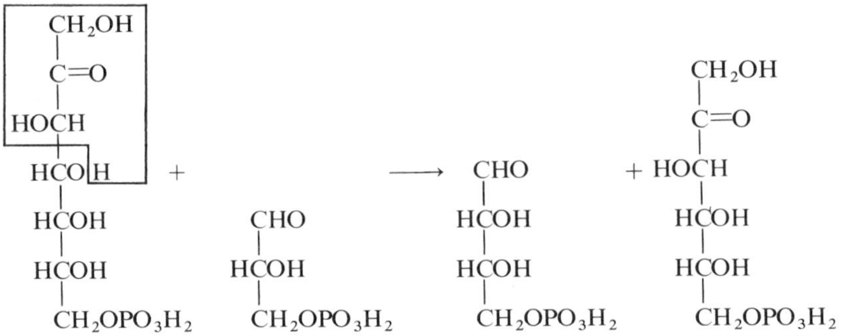

Both enzymes together carry out the conversion of pentose-5-phosphates into fructose-6-phosphate and glyceraldehyde-3-phosphate according to the equation:

$$2 \text{ xylulose-5-P} + \text{ribose-5-P} \rightleftharpoons 2 \text{ fructose-6-P} + \text{glyceraldehyde-3-P}$$

Consequently, any pentose phosphate in excess can be conveniently channeled into the Embden–Meyerhof pathway. Since the reactions involved are reversible, the synthesis of pentose phosphates from hexose phosphates is also possible under conditions under which the formation of $NADPH_2$ by the oxidative portion of the cycle is not desired. Moreover, the reactions of the pentose phosphate pathway provide the cells with erythrose-4-phosphate, which, for instance, is required for aromatic amino acids biosynthesis.

E. coli channels about 28% of the glucose that is degraded into the

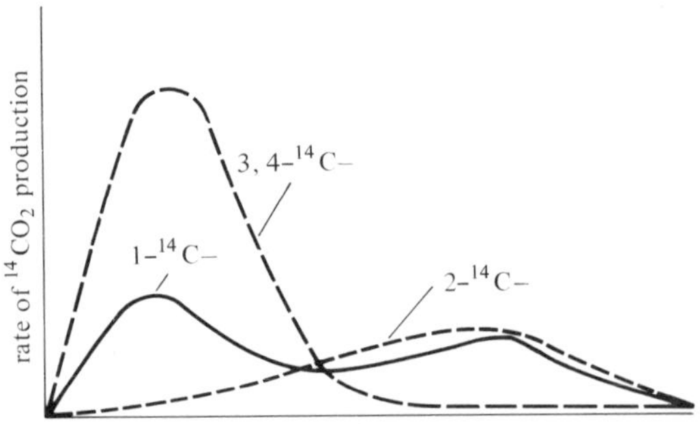

Figure 3.10. Formation of radioactive CO_2 by *E. coli* cells from glucose labeled with ^{14}C in various positions. The amount of radioactive CO_2 formed per hour is plotted against the time. [C. H. Wang, I. Stern, C. M. Gilmour, S. Klungsoyr, D. J. Reed, J. J. Bialy, B. E. Christensen, and V. H. Cheldelin: *J. Bacteriol.* **76**, 207–216 (1958).]

oxidative pentose phosphate pathway and 72% into the Embden–Meyerhof pathway. This can be calculated from **radiorespirometric experiments** similar to that shown in Figure 3.10, which demonstrates the rate of release of radioactive carbon dioxide from cultures of *E. coli* growing with radioactive glucose labeled in various positions. As can be seen, carbon atoms 3 and 4 are the preferential sources of CO_2; in the degradation of glucose by the Embden–Meyerhof pathway C_3 and C_4 of glucose are converted to carboxyl groups of pyruvate and released as CO_2 during acetyl-CoA formation. Radioactivity from 1-^{14}C-glucose appears in CO_2 when 6-phosphogluconate is oxidized to ribulose-5-phosphate (first peak) and when acetyl-CoA is oxidized via the tricarboxylic acid cycle (second peak). The latter process gives also rise to $^{14}CO_2$ production from 2-^{14}C-glucose.

VI.   Ribonucleotides and Deoxyribonucleotides

Ribonucleotides consist of a purine or pyrimidine base, ribose, and a phosphate, pyrophosphate, or triphosphate group:

$$
\begin{array}{ll}
\text{purine} & \underline{\hspace{2em}} \\
\text{pyrimidine} & \underline{\hspace{2em}}
\end{array}
\quad \text{ribose}
\begin{array}{l}
\underline{\hspace{2em}} \; \text{P} \\
\underline{\hspace{2em}} \; \text{P - P} \\
\underline{\hspace{2em}} \; \text{P - P - P}
\end{array}
$$

Table 3.4 summarizes the structures of important purines and pyrimidines and the corresponding nucleotides.

A purine or pyrimidine base attached to ribose is called a **ribonucleoside** (trivial names: adenosine, guanosine, cytidine, and uridine). Esterification of the ribonucleosides with phosphoric acid yields the corresponding ribonucleoside monophosphates. Their linkage to a second and a third phosphate residue results in the ribonucleoside di- and triphosphates. All these phosphate esters are referred to as **nucleotides**.

The precursors of the **pyrimidine nucleotides** are carbamyl phosphate and aspartate. The enzyme **aspartate transcarbamylase** condenses these compounds to yield carbamyl aspartate, which undergoes cyclization to give 4,5-dihydroorotate. Dehydrogenation then leads to orotate, the first intermediate containing the pyrimidine ring (Figure 3.11).

Before orotate is converted to one of the important physiological pyrimidine bases it is linked to ribose-5-phosphate to form the corresponding ribonucleotide. Ribose-5-phosphate itself cannot function as substrate in this condensation reaction; it must be activated by conversion into **5-phosphoribosyl-1-pyrophosphate (PRPP)** (Figure 3.12). The reaction of 5-phosphoribosyl-1-pyrophosphate with orotate then yields orotidine monophosphate, which is subsequently decarboxylated to uridine monophosphate (**UMP**).

Table 3.4. Important pyrimidines, purines, and corresponding nucleotides

base	nucleotide

uracil   uracil-ribose- Ⓟ
                          uridine monophosphate (UMP)

cytosine   cytosine-ribose- Ⓟ
                          cytidine monophosphate (CMP)

thymine   thymine-ribose- Ⓟ
                          thymidine monophosphate (TMP)

adenine   adenine-ribose- Ⓟ
                          adenosine monophosphate (AMP)

guanine   guanine-ribose- Ⓟ
                          guanosine monophosphate (GMP)

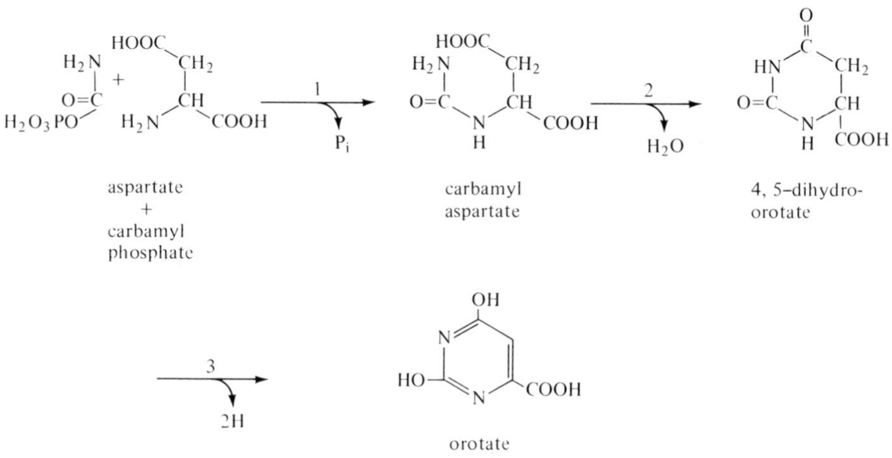

Figure 3.11. Biosynthesis of orotate from carbamyl phosphate and aspartate. 1, aspartate transcarbamylase; 2, dihydroorotase; 3, dihydroorotate dehydrogenase.

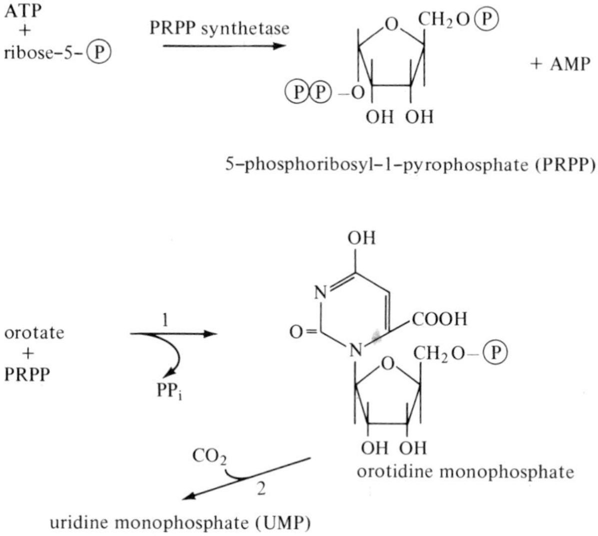

Figure 3.12. Formation of UMP from orotate. 1, orotidylate pyrophosphorylase; 2, orotidylate decarboxylase.

UMP can be converted into UTP by kinase reactions; a subsequent amination reaction yields cytidine triphosphate (**CTP**):

$$UMP \xrightarrow[ADP]{ATP} UDP \xrightarrow[ADP]{ATP} UTP$$

The synthesis of the **purine nucleotides** is more complicated. Starting with PRPP an imidazole nucleotide is formed (Figure 3.13).

The three pyrimidine ring atoms still required for the formation of the purine ring from 5-aminoimidazole are furnished by bicarbonate, aspartate, and formyl H_4-folate. Ring closure then yields inosinic acid (Figure 3.14). From IMP a number of additional reactions lead either to **AMP** or **GMP**, and the enzyme adenylate kinase (myokinase in animal tissues) then converts AMP into ADP:

$$AMP + ATP \underset{\longleftarrow}{\overset{\text{adenylate kinase}}{\rightleftharpoons}} 2ADP$$

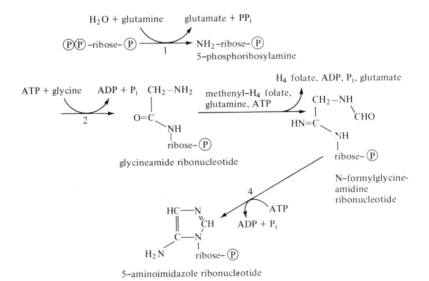

Figure 3.13. First reactions of purine nucleotide synthesis. 1, **PRPP** amidotransferase; 2, phosphoribosyl-glycineamide synthetase; 3, phosphoribosyl-glycineamide formyl-transferase + phosphoribosyl-formylglycineamide synthetase; 4, phosphoribosyl-amino-imidazole synthetase.

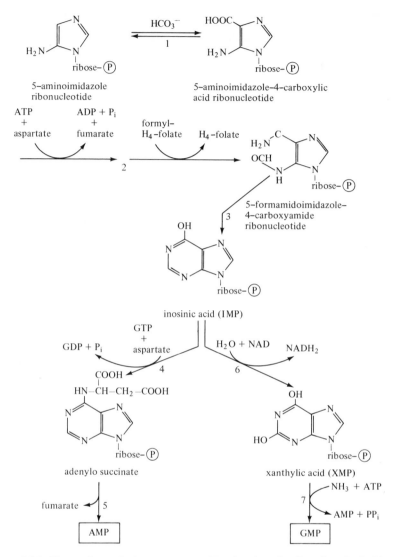

Figure 3.14. Formation of AMP and GMP. 1, phosphoribosyl-aminoimidazole carboxylase; 2, phosphoribosyl-aminoimidazole succinocarboxamide synthetase + adenylosuccinate lyase + phosphoribosyl-aminoimidazole carboxamide formyltransferase; 3, IMP cyclohydrolase; 4, adenylosuccinate synthetase; 5, adenylosuccinate lyase; 6, IMP dehydrogenase; 7, GMP synthetase.

Finally, ATP can be formed by electron transport or substrate-level phosphorylation. Additional nucleoside monophosphate kinases are present in *E. coli* to phosphorylate GMP with ATP, the ultimate source of high-energy phosphate.

The role of purine and pyrimidine nucleotides in cell metabolism is diverse, Numerous enzymatic reactions depend on ATP, GTP, UTP, or CTP as sources of either high-energy phosphate or groups with a high potential of group transfer (e.g., ADP in ADP-glucose). Ribonucleotides are the precursors of the deoxynucleotides and, therefore, serve as substrates for both RNA and DNA synthesis.

A. Synthesis of deoxyribonucleotides

Reduction of ribonucleotides to deoxyribonucleotides takes place in *E. coli* at the diphosphate level:

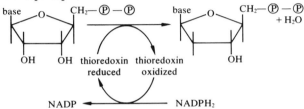

The reducing agent in this reaction is thioredoxin, a flavoprotein; its reduced form is regenerated with $NADPH_2$.

Four nucleoside diphosphates are reduced according to this scheme:

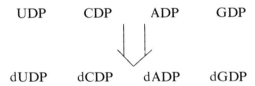

UDP   CDP   ADP   GDP

dUDP   dCDP   dADP   dGDP

Deoxyuridinephosphate is not a major constituent of DNA. However, dUDP is the precursor of thymine-containing deoxynucleotides; it is hydrolyzed to dUMP and then methylated:

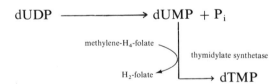

$$dUDP \longrightarrow dUMP + P_i$$

methylene-H_4-folate

thymidylate synthetase

H_2-folate

$\longrightarrow dTMP$

In the methylation of dUMP to dTMP, methylene-H_4-folate plays a unique role; it is not only the donor of the C_1-unit but serves also as reducing agent. There are kinases present in *E. coli* that catalyze the formation of the corresponding deoxynucleosidetriphosphate with ATP as donor of high-energy phosphate.

VII.   Biosynthesis of Lipids

Bacteria do not accumulate lipids as reserve material but they contain considerable amounts of lipids as constituents of their membrane systems, especially phospholipids and glycolipids. The general structure of these lipids is shown in Figure 3.15.

```
CH₂–O–CO–R
|
CH–O–CO–R              lipid (neutral fats)
|
CH₂–O–CO–R
```

```
CH₂–O–CO–R              phospholipid
|
CH–O–CO–R
|        O
|        ||          /CH₃
CH₂–O–P–O–CH₂–CH₂–N⁺←–CH₃
         |          \CH₃
         O⁻
```

```
CH₂–O–CO–R              glycolipid
|
CH–O–CO–R
|
|              CH₂OH
|          O—|
CH₂–O—⟨OH OH|
           |—|  OH
            OH
```

Figure 3.15. General structure of lipids, phospholipids, and glycolipids. R, Carbon chains of fatty acids esterified with glycerol. Choline (shown here) is one of the alcohols that can be present in phospholipids; others are ethanolamine, serine, and inositol. In the glycolipid shown D-mannose is the sugar moiety. Other glycolipids contain galactose, glucose, or oligosaccharides.

   The following is an outline of the synthesis of the major constituents of lipids: fatty acids, glycerol, and choline.

A.   Fatty acids

Most of the fatty acids occurring in lipids contain 16 or 18 carbon atoms; they are saturated or have one or more double bonds. The precursor of fatty acids is acetyl-CoA. However, chain elongation is not achieved by condensation of two acetyl-CoA molecules followed by further condensation of the C_4-compound with acetyl-CoA, a reaction sequence accomplished by clostridia when forming butyrate and caproate. Two variations are important:

1. CoA-derivatives are not substrates of the enzymes involved in fatty acid synthesis. Instead *E. coli* employs an **acyl carrier protein (ACP)** of a molecular weight of 10,000; its prosthetic group is 4′-phosphopantetheine, and it thus resembles coenzyme A (Figure 3.16).

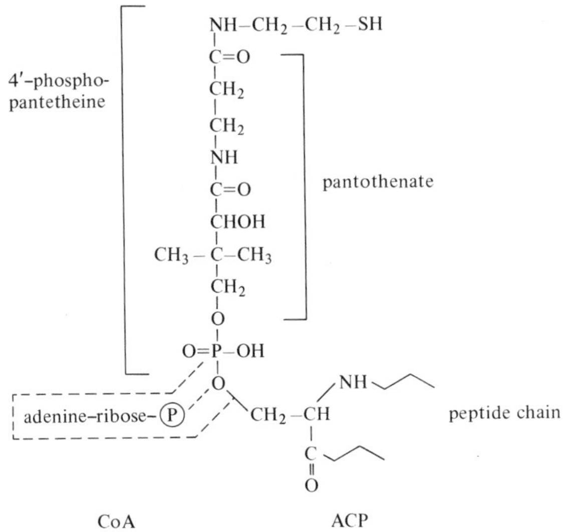

Figure 3.16. Structure of the prosthetic group of ACP and CoA. In ACP the 4′-phosphopantetheine is linked to the peptide chain via a serine residue.

The first reaction in fatty acid synthesis is the formation of acetyl-ACP:

$$\text{acetyl-CoA} + \text{ACP} \xrightleftharpoons[\hspace{2cm}]{\text{acetyl transacetylase}} \text{acetyl-ACP} + \text{CoA}$$

2. Acetyl-ACP functions as primer in fatty acid synthesis and the C_2-units are added to this primer in the form of malonyl-ACP. The latter is synthesized from acetyl-CoA in two steps:

$$\text{CH}_3\text{—CO—CoA} + \text{ATP} + \text{CO}_2 \xrightleftharpoons[\hspace{2cm}]{\text{acetyl-CoA carboxylase}}$$

$$\begin{array}{l} \text{CH}_2\text{—CO—CoA} + \text{ADP} + \text{P}_i \\ | \\ \text{COOH} \end{array}$$

$$\text{malonyl-CoA} + \text{ACP} \xrightleftharpoons[\hspace{1cm}]{\text{malonyl transacylase}} \text{malonyl-ACP} + \text{CoA}$$

Acetyl-CoA carboxylase is a biotin-containing enzyme.

Four enzymes are required to elongate an acyl-ACP by two carbon atoms. First acetyl-ACP reacts with malonyl-ACP to form acetoacetyl-ACP, which is then reduced to D(-)-β-hydroxybutyryl-ACP. Dehydration yields crotonyl-ACP and subsequent reduction results in butyryl-ACP (Figure 3.17).

The names of the enzymes shown in Figure 3.17 suggest that they may be relatively unspecific with respect to the chain length of their substrates. Indeed, they also catalyze the conversion of butyryl-ACP and malonyl-ACP to caproyl-ACP as well as all subsequent reactions until palmityl-ACP is formed. The latter is the predominant saturated acyl-ACP produced in

Table 3.5. Activity of 3-ketoacyl-ACP synthetase with various acyl-ACP compounds[a]

substrate	velocity of reaction V_{max} (μmol product/min/mg)
acetyl-ACP	2.8
decanoyl-ACP	2.8
dodecanoyl-ACP	0.97
tetradecanoyl-ACP	0.31
hexdecanoyl-ACP	zero
cis-5-dodecenoyl-ACP	1.7
cis-9-hexadecenoyl-ACP	0.37
cis-11-octadecenoyl-ACP	zero

[a]D. J. Prescott and P. R. Vagelos, *Adv. Enzymol.* **36**, 269–311 (1972).

E. coli. This is because the 3-keto-acyl-ACP synthetase is active with substrates having a maximum chain length of C_{14} (Table 3.5). Thus, with regard to the saturated acyl derivatives, palmityl-ACP is the final product.

3–ketoacyl–ACP synthetase
acetyl–ACP + malonyl–ACP $\rightleftharpoons$ acetoacetyl–ACP + CO_2 + ACP

3–ketoacyl–ACP reductase
acetoacetyl–ACP + $NADPH_2$ $\rightleftharpoons$ β–hydroxybutyryl–ACP + NADP

β–hydroxyacyl–ACP dehydratase
β–hydroxybutyryl–ACP $\rightleftharpoons$ crotonyl–ACP + H_2O

enoyl–ACP reductase
crotonyl–ACP + $NADPH_2$ $\rightleftharpoons$ butyryl–ACP + NADP

Figure 3.17. Reactions involved in the formation of butyryl-ACP from acetyl-ACP and malonyl-ACP.

Some **unsaturated fatty acids** are also important constituents of the phospholipids. Those predominating in *E. coli* are **palmitoleate** (*cis*-9-hexadecenoate) and **cis-vaccenate** (*cis*-11-octadecenoate). Their synthesis involves the enzyme systems discussed above and it is evident from Table 3.5 that the synthetase is still active with the unsaturated C_{16}-ACP so that *cis*-vaccenyl-ACP can be formed. The branch point of saturated and unsaturated fatty acid synthesis in *E. coli* is at β-hydroxydecanoyl-ACP.

$$CH_3\!-\!(CH_2)_5\!-\!CH_2\!-\!CHOH\!-\!CH_2\!-\!CO\!-\!ACP$$

β-hydroxydecanoyl-ACP

special
dehydratase

3-hydroxyacyl-ACP
dehydratase

$$CH_3\!-\!(CH_2)_5\!-\!\overset{\overset{\textstyle H}{|}}{C}\!=\!\overset{\overset{\textstyle H}{|}}{C}\!-\!CH_2\!-\!CO\!-\!ACP \qquad CH_3\!-\!(CH_2)_5\!-\!CH_2\!-\!\overset{\overset{\textstyle H}{|}}{C}\!=\!\underset{\underset{\textstyle H}{|}}{C}\!-\!CO\!-\!ACP$$

unsaturated acids   saturated acids

A special dehydratase removes water to yield a compound with a *cis*-double bond between carbon atoms 3 and 4; the resulting compound does not function as a substrate for enoyl-ACP reductase but is subject to further elongation reactions, which yield C_{16} and C_{18} mono-unsaturated acyl-ACP.

It should be mentioned that all the enzymes involved in fatty acid synthesis in *E. coli* are soluble and readily separable from one another *in vitro*. In higher organisms all the reactions leading from acetyl-CoA and malonyl-CoA to long-chain fatty acids are catalyzed by a multienzyme complex: fatty acid synthetase.

B. Phosphatidic acid

The principal substrates for the formation of phosphatidic acids are glycerol-3-phosphate and acyl-ACP. The former is readily available from dihydroxyacetonephosphate—an intermediate of the Embden–Meyerhof pathway:

glycerol-3-phosphate
dehydrogenase

$$HOCH_2\!-\!CO\!-\!CH_2O\textcircled{P} + NADPH_2 \rightleftharpoons$$
$$HOCH_2\!-\!CHOH\!-\!CH_2O\textcircled{P} + NADP$$

Phosphatidic acids are then synthesized as follows:

$$\begin{array}{l} CH_2OH \\ | \\ CHOH \\ | \\ CH_2O\,\textcircled{P} \end{array} \quad + \ 2R\!-\!CO\!-\!ACP \quad \xrightarrow[\substack{\downarrow \\ 2\ ACP}]{\substack{\text{glycerol phosphate} \\ \text{acyltransferase}}} \quad \begin{array}{l} CH_2O\!-\!CO\!-\!R \\ | \\ CHO\!-\!CO\!-\!R \\ | \\ CH_2O\,\textcircled{P} \end{array}$$

phosphatidic acid

After hydrolytic removal of phosphate from the phosphatidic acid, neutral fats may be formed by reaction with a third acyl-ACP. Most of the phosphatidic acid, however, is used for the synthesis of phospholipids.

C. Phospholipids

The phosphate group of phosphatidic acid is prepared for esterification with an alcohol by the reaction with CTP:

$$
\begin{array}{l}
CH_2O{-}CO{-}R \\
 \\
CHO{-}CO{-}R \quad + CTP \\
 \\
\overset{\displaystyle O}{\underset{\displaystyle OH}{CH_2O{-}\overset{\parallel}{P}{-}OH}}
\end{array}
\qquad
\begin{array}{l}
CH_2O{-}CO{-}R \\
 \\
CHO{-}CO{-}R \\
 \\
CH_2O{-}\overset{\displaystyle O}{\underset{\displaystyle OH}{\overset{\parallel}{P}}}{-}O{-}\overset{\displaystyle O}{\underset{\displaystyle OH}{\overset{\parallel}{P}}}{-}O\text{-cytidine}
\end{array}
\qquad + PP_i
$$

phosphatidic acid   CDP-diacylglycerol

Specific enzymes then catalyze the displacement of CMP by alcohols, like serine, inositol, and glycerol.

$$CDP\text{-diacylglycerol} + serine \longrightarrow phosphatidyl\text{-serine} + CMP$$

Decarboxylation of phosphatidyl-serine yields phosphatidyl-ethanolamine, which can be methylated with S-adenosyl-methionine to yield phosphatidyl-choline.

$$
\begin{array}{l}
CH_2O{-}CO{-}R \\
 \\
CHO{-}CO{-}R \\
 \\
CH_2O{-}\overset{\displaystyle O^-}{\underset{\displaystyle O}{\overset{|}{\underset{\parallel}{P}}}}{-}O{-}CH_2{-}CH_2{-}NH_2
\end{array}
\quad + \quad 3 \quad
\begin{array}{l}
COO^- \\
| \\
HC{-}NH_2 \\
| \\
CH_2 \\
| \\
CH_2 \\
| \\
H_3C{-}S\text{-adenosine} \\
+
\end{array}
$$

phosphatidyl-ethanolamine   S-adenosyl-methionine

$$
\begin{array}{l}
CH_2O{-}CO{-}R \\
 \\
CHO{-}CO{-}R \\
 \\
CH_2{-}O{-}\overset{\displaystyle O^-}{\underset{\displaystyle O}{\overset{|}{\underset{\parallel}{P}}}}{-}O{-}CH_2{-}CH_2{-}\overset{+}{N}\underset{\diagdown CH_3}{\overset{\diagup CH_3}{-CH_3}}
\end{array}
\qquad + 3 \;\; \begin{array}{l} S\text{-adenosyl-} \\ homocysteine + 2H^+ \end{array}
$$

phosphatidyl-choline

Analogous pathways are employed for the synthesis of other phospholipids.

VIII.   Formation of Carbohydrates

When glucose serves as growth substrate for *E. coli* a number of hexose and pentose phosphates are intermediates in the breakdown of this substrate by the Embden–Meyerhof pathway and the oxidative pentose phosphate pathway: glucose-6-phosphate, fructose-6-phosphate, fructose-1,6-bisphosphate, ribulose-5-phosphate, xylulose-5-phosphate, and ribose-5-phosphate. For the synthesis of **galactose** esters, glucose is first linked to UDP by the following reactions:

$$\text{glucose-6-phosphate} \underset{}{\overset{\text{phosphoglucomutase}}{\rightleftharpoons}} \text{glucose-1-phosphate}$$

$$\text{glucose-1-phosphate} + \text{UTP} \underset{}{\overset{\substack{\text{UDP-glucose}\\\text{pyrophosphorylase}}}{\rightleftharpoons}} \text{UDP-glucose} + \text{PP}_i$$

A change of the configuration at carbon atom 4 of the glucose moiety is then accomplished by a specific epimerase, which has an absolute requirement for NAD. A 4-ketoglucose derivative is an intermediate in this reaction:

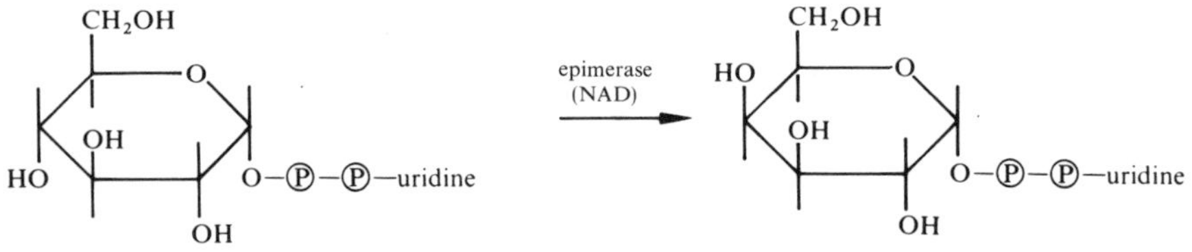

UDP-glucose   →   UDP-galactose

UDP-galactose is a precursor of *E. coli* wall lipopolysaccharide. Other precursors of this important structural element are **hexose amines**; they originate from fructose-6-phosphate, which is converted into glucosamine-6-phosphate with glutamine as NH_2-donor:

$$\text{D-fructose-6-phosphate} \xrightarrow[\substack{\text{L-glutamine} \quad \text{L-glutamate}}]{\text{aminotransferase}} \text{D-glucosamine-6-phosphate}$$

Other hexose amines are derived from glucosamine-6-phosphate. **UDP-N-acetylmuramic acid**, precursor of peptidoglycan (murein), is also synthesized from this compound (Figure 3.18). First the phosphate group is transferred from position 6 to position 1. Subsequent reaction with acetyl-CoA yields N-acetylglucosamine-1-phosphate, which is then linked to UDP. In the last step UDP-N-acetylglucosamine reacts with PEP to yield the 3-lactyl derivative, which is called UDP-N-acetylmuramic acid.

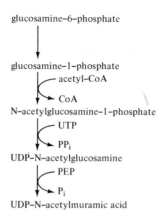

glucosamine–6–phosphate

glucosamine–1–phosphate
    acetyl–CoA
    CoA
N–acetylglucosamine–1–phosphate
    UTP
    PP_i
UDP–N–acetylglucosamine
    PEP
    P_i
UDP–N–acetylmuramic acid

Figure 3.18. Formation of UDP-N-acetylmuramic acid from glucosamine-6-phosphate.

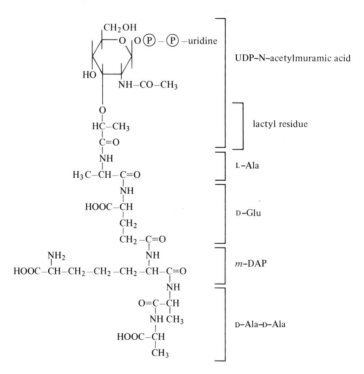

Figure 3.19. UDP-N-acetylmuramyl-pentapeptide, a building block of the peptidoglycan of *E. coli*.

The lactyl residue of UDP-N-acetylmuramic acid is very important; the building block of the **peptidoglycan** (murein) of the cell wall is a **muramyl-pentapeptide**, and the peptide chain is attached to the carboxyl group of this lactyl residue (Figure 3.19). The *E. coli* pentapeptide is synthesized from UDP-N-acetylmuramic acid by subsequent reaction with L-alanine + ATP, D-glutamate + ATP, *meso*-diaminopimelate + ATP, and D-alanyl-D-alanine + ATP. The formation of D-alanyl-D-alanine from D-alanine is specifically inhibited by the antibiotic cycloserine:

$$2\text{-D-alanine} + \text{ATP} \xrightarrow{\quad\text{cycloserine}\quad\!\!\!\!\!\!\!\!|\!|\!\!\!\!\!\!\!\!\quad} \text{D-alanyl-D-alanine} + \text{ADP} + \text{P}_i$$

IX.   Synthesis of Polymers

It has been mentioned that about 97% of the cellular material are macromolecules. Three groups can be distinguished: **lipids, periodic macromolecules** such as peptidoglycan and polysaccharides, and **informational macromolecules** such as nucleic acids and proteins.

A.   Lipids

Lipids are not true macromolecules, as the monomers are not linked to one another by covalent bonds. However, in an aqueous environment phospholipid molecules like phosphatidyl-choline associate in such a way that a double-layered structure is formed, as shown in Figure 3.20(a). This structure is stabilized by interaction of the hydrophobic "tails" and by hydrogen bonding between the hydrophylic "heads" of the phospholipid. Together with proteins, lipids associate to membranes; their general structure is shown in Figure 3.20(b). Membranes contain up to 60% of proteins; some are embedded in the lipid layers and have specific functions in the various transport processes. A membrane of this general structure surrounds the cytoplasm of the bacterial cell and is called cell or cytoplasmic membrane.

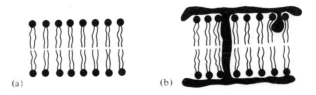

(a)   (b)

Figure 3.20. Arrangement of phospholipid molecules in membranes. (a) double layer of phospholipid molecules (○, hydrophylic head; ≈, hydrophobic tails); (b) model of a membrane containing phospholipids and proteins.

B. Periodic macromolecules

Unlike the informational macromolecules, the periodic molecules are not synthesized along a template. Rather, they are formed by specific enzymes from one (glycogen), two (peptidoglycan), or several (lipopolysaccharides) types of building blocks.

1. Glycogen. When *E. coli* grows on glucose it stores some of the glucose taken up into the cell as glycogen. Synthesis of glycogen involves two steps: activation of glucose and transfer of the glucose moiety to a primer molecule. The activated glucose derivative is **ADP-glucose**, and it is formed from glucose-1-phosphate by **ADP-glucose pyrophosphorylase**:

$$\text{glucose-1-P} + \text{ATP} \longrightarrow \text{ADP-glucose} + \text{PP}_i$$

There is then a transfer of the glucose moiety to the nonreducing end of an oligosaccharide (primer), which must consist of more than four glucose residues (Figure 3.21).

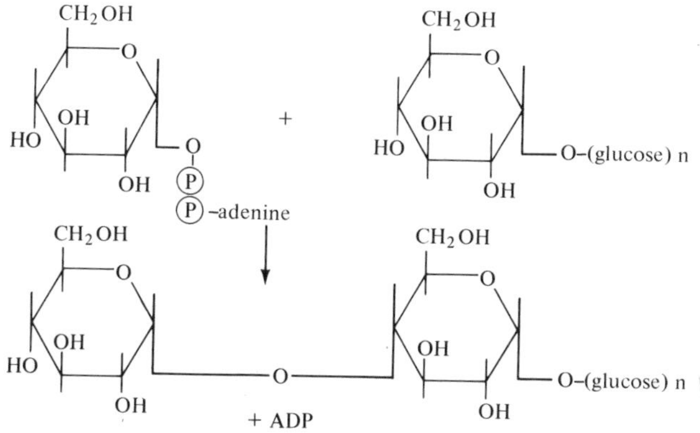

Figure 3.21. Elongation of a polysaccharide chain of glycogen. The enzyme is called glycogen synthetase.

E. coli glycogen is a highly branched polysaccharide and contains, in addition to the α-$(1\rightarrow4)$ linkages between glucose molecules, a large number of α-$(1\rightarrow6)$ linkages. The latter are formed by **transglucosylation**; an oligosaccharide is transferred from the nonreducing end to the OH-group at C_6 of a glucose residue in the chain (Figure 3.22).

It should be mentioned that in animals UDP-glucose is the substrate for glycogen synthesis.

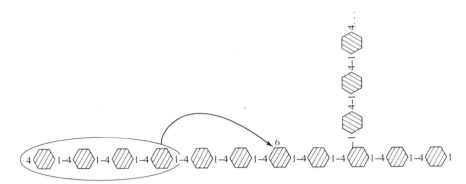

Figure 3.22. Branching by transglucosylation.

2. Cell wall. Figure 3.23 presents a schematic view of the macromolecular structures surrounding the cytoplasm of *E. coli*. It can be seen that the cell membrane is covered by a rather thin **peptidoglycan layer**, as is characteristic of Gram-negative bacteria. The final layer is the **outer membrane** and it consists of lipopolysaccharides, lipids, and proteins.

Figure 3.23. Surface layers of *E. coli* in thin section. CM, cell membrane; OM, outer membrane; PG, peptidoglycan.

Peptidoglycan (murein) is synthesized from two building blocks: UDP-N-acetylmuramyl-pentapeptide and UDP-N-acetylglucosamine. The formation of these compounds in the cytoplasm was outlined. As shown in Figure 3.24 N-acetylmuramyl-pentapeptide is first transferred to a membrane lipid carrier—**undecaprenyl phosphate (bactoprenol)**:

$$CH_3$$
$$\diagdown$$
$$C=CH-CH_2-(CH_2-\overset{\overset{\textstyle CH_3}{|}}{C}=CH-CH_2)_9-CH_2-\overset{\overset{\textstyle CH_3}{|}}{C}=CH-CH_2O\textcircled{P}$$
$$\diagup$$
$$CH_3$$

bactoprenol or undecaprenyl phosphate

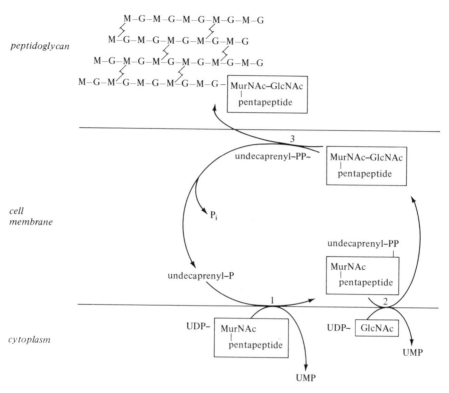

Figure 3.24. Reactions involved in the elongation of a peptidoglycan chain by a disaccharide peptide. GlcNAc (G), N-acetylglucosamine; MurNAc (M), N-acetylmuramic acid; 4, peptide bridges. 1, transfer of MurNAc-pentapeptide to undecaprenyl-P; 2, formation of disaccharide-pentapeptide; 3, transfer to peptidoglycan chain.

It is then linked to N-acetylglucosamine. This disaccharide pentapeptide is then transported by the lipid carrier to the extracellular site of the cell membrane and used as a building block in peptidoglycan chain elongation. Presumably *E. coli* peptidoglycan consists of a monomolecular layer of chains with the polysaccharide backbone facing the outer membrane and the peptide residues facing the cell membrane. Cross-linkage between adjacent chains is brought about by the closure of bridges between some of the peptide units. This is accomplished by a transpeptidation reaction. As shown in Figure 3.25 the peptide bond between the two D-alanine residues is replaced by a peptide bond between D-alanine and a *m*-diaminopimelate residue of an adjacent chain. The D-alanine released is transported back into the cytoplasm.

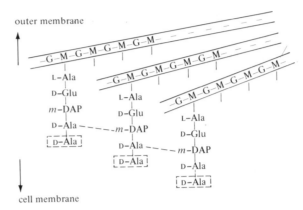

Figure 3.25. Cross-linkage by transpeptidation. D-Alanine is released from one peptide unit and a peptide bond is formed between the remaining D-alanine residue and the free amino group of the diaminopimelate residue of an adjacent peptide unit. Terminal D-alanine residues, which do not participate in transpeptidation reactions, are removed by a D-alanine carboxypeptidase.

Cross-linkage by transpeptidation outside the cytoplasm seems to be advantageous since it uses the energy of the peptide bond and, therefore, does not require ATP. It also is apparent that for cross-bridging, an amino acid with a free functional group must be present in position 3 of the peptide unit. In *E. coli* peptidoglycan this position is occupied by diaminopimelate while the peptidoglycan of other microorganisms contains lysine or ornithine (Chapter 5, Section VIII).

Several antibiotics interfere with peptidoglycan synthesis. **Penicillin** and **cephalosporin** inhibit the transpeptidation reaction. Since these cross-bridging reactions occur in growing, but not in resting, cells it is understandable that penicillin and cephalosporin preferentially kill growing cells. Penicillin is not a very potent drug against Gram-negative bacteria like *E. coli* because in order to reach the peptidoglycan layer the penicillin must penetrate the rather thick lipopolysaccharide layer and other components of the outer membrane. Gram-positive bacteria are more severely affected by penicillin than are Gram-negative bacteria.

3. Outer membrane layer. This layer consists of about 50% lipopolysaccharide, 35% phospholipids, and 15% protein. The arrangement of these components is shown schematically in Figure 3.26.

Lipid A contains fatty acids different from those present in phospholipids; most notably 3-hydroxytetradecanoic acid is found. Six molecules of this acid are linked to the glucosamine disaccharide of lipid A through ester and amide linkages. In addition, the disaccharide carries a phosphate group and pyrophosphorylethanolamine (Fig. 3.27). Lipid A is interspersed among phospholipid molecules and carries the **core oligosaccharide**. The latter contains two unusual components, heptose and 3-deoxy-D-*manno*-octulo-

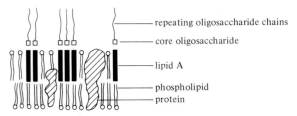

Figure 3.26. Arrangement of components of the outer membrane layer.

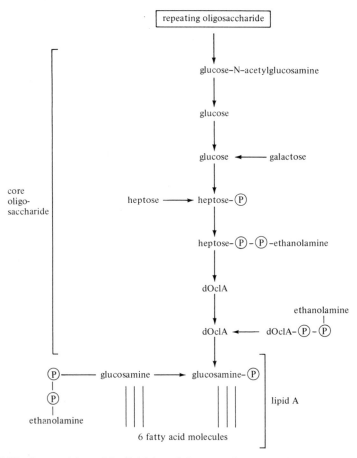

Figure 3.27. Composition of the lipid A and the core oligosaccharide of *E. coli*. dOclA, 3-deoxy-D-*manno*-octulosonate (formerly abbreviated as KDO).

sonate (formerly called 2-keto-3-deoxyoctonate, KDO). Present in the core oligosaccharide are also glucose, galactose and N-acetylglucosamine.

CH$_2$OH   CH$_2$OH
| |
HOCH   HOCH
                                        COOH
OH   OH   OH
HO   OH   OH

heptose   3-deoxy-D-*manno*-octulosonate(dOcl A)

Synthesis of the core oligosaccharide occurs stepwise. First UDP-derivatives of the appropriate sugars travel by unknown mechanisms to the lipopolysaccharide area and are added to the growing chain by specific transferases. The composition of the core oligosaccharide is almost identical in all enterobacteria. However, great differences exist in the composition of the **repeating oligosaccharide chains**. Usually 4 to 6 sugars form a repeating unit ; within one species, such as *E. coli* or especially *Salmonella typhimurium*, there is a large variation in the nature of these sugars and their sequence. One example of a repeating unit found in *E. coli* is shown in Figure 3.28. The repeating oligosaccharide chains are primarily responsible for the somatic (o) antigenic specificity of Gram-negative bacteria.

Biosynthesis of the repeating units proceeds by reactions analogous to those of peptidoglycan formation. The enzymes involved are membrane-

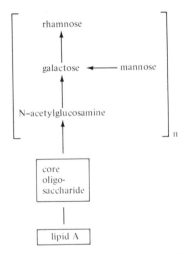

Figure 3.28. Composition of a repeating oligosaccharide chain of *E. coli*.

bound and undecaprenyl phosphate serves as lipid carrier. Nucleoside diphosphate-activated sugars are transferred to undecaprenyl phosphate in the right order to form an undecaprenyl-PP-oligosaccharide. At the outer surface of the cell membrane these units are polymerized and then transferred to the core oligosaccharide.

C. Synthesis of informational macromolecules

The macromolecules we have thus far discussed are composed of repeating sequences of building blocks. Glycogen, for instance, consists of glucose molecules connected to one another by α-$(1\rightarrow4)$ or α-$(1\rightarrow6)$ linkages. Peptidoglycan is a polymer with an alternating sequence of N-acetylglucosamine and N-acetylmuramic acid. The latter bears a tetrapeptide chain, which for a particular microorganism, is always composed of the same four amino acids. It is then understandable that these macromolecules can be synthesized by a small number of specific enzymes and that there is no requirement for a template to bring the building blocks together in the appropriate sequence. This becomes different when we turn to the synthesis of nucleic acids and proteins. Here, regular repeating units are absent and the synthesis of these macromolecules must proceed at a template containing the necessary information. With the exception of certain viruses the same hierarchy in template function is established in all organisms: DNA serves as template for DNA and RNA synthesis; RNA serves as template for protein synthesis. The processes involved will be discussed here only briefly.

1. DNA and RNA synthesis. DNA is a double helix (Watson–Crick model). The two polynucleotide strands are held together by hydrogen bonds, which can be most effectively formed only between two pairs of the four bases present in DNA, between adenine and thymine and between guanine and cytosine. Figure 3.29 shows a short piece of a DNA double helix and Figure 3.30 a duplex of two pentanucleotides. It can be seen that within one strand the nucleosides are connected to one another by phosphate bridges between the 3' and 5' carbons of deoxyribose. Both strands run antiparallel, i.e., in the left-hand strand the chain proceeds from a terminal 5'-phosphate group to a 3'-hydroxyl group, while the opposite is true in the right-hand strand.

Figure 3.31 shows the first two base pairs in more detail. The hydrogen bonds allow pairing between A and T and between G and C so perfectly that other combinations of the bases are not possible. Since the genetic information stored in DNA is in the form of defined base sequences, the following mechanism lends itself to the duplication of this information: The two strands become separated; two new strands are formed along the parental strands, and the nature of each nucleotide added to the growing chain is determined by the nature of the base at a similar position in the parental strand. The above duplication is catalyzed by DNA polymerase and the process is called **replication** (illustrated in Figure 3.32).

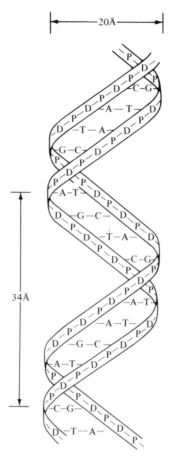

Figure 3.29. Double-helix structure of DNA. D, deoxyribose; P, phosphate ester bridge; A, adenine; C, cytosine; G, guanine; T, thymine.

DNA polymerase was discovered in *E. coli* by A. Kornberg and associates in 1955. The enzyme catalyzes DNA formation from the four deoxyribonucleoside triphosphates only when a template is present. As is indicated in Figure 3.32 the monomers are added to the 3'-hydroxyl group of the growing strand. Because of this substrate specificity the question then arises as to how the other parental strand can be replicated at the same time. There are indications that the DNA polymerase—after a number of polymerization steps at parental strand A—jumps to the other parental strand (B) and replicates it going backward. A second round of replication would then start at parental strand A again. The final result of replication is the formation of two double-stranded DNA molecules from one, each containing one parental and one newly synthesized strand.

E. *coli* contains three DNA polymerases. The enzyme first discovered by Kornberg is probably primarily a repair enzyme, and the other two enzymes

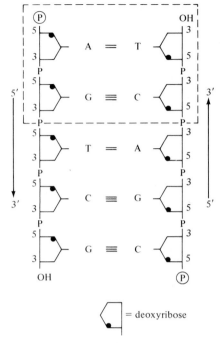

Figure 3.30. A duplex of two pentanucleotides.

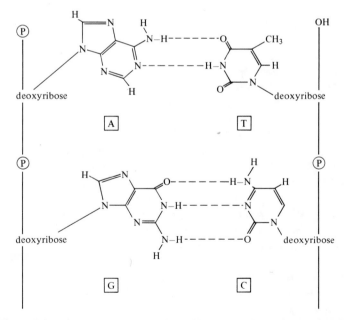

Figure 3.31. AT and GC base pairs held together by hydrogen bonds.

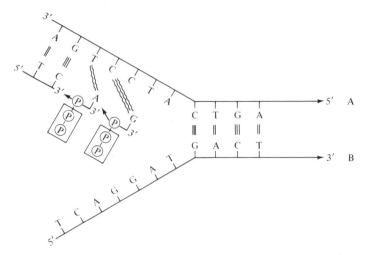

Figure 3.32. DNA replication. The polymerase adds the incoming nucleotide at the 3'-hydroxyl group, so that the strand grows in the 5'→3' direction. Pyrophosphate is released during polymerization.

are mainly involved principally in DNA replication. They all catalyze the same overall reaction:

$$\text{n deoxynucleoside triphosphate} \xrightarrow{\text{template}} \text{DNA}_n + \text{n PP}_i$$

The pyrophosphate is hydrolyzed to phosphate by a pyrophosphatase. Consequently, DNA formation costs the cell two ATP per monomer. Since the DNA content of cells is approximately 3%, only a small percentage of the ATP required for growth is invested in DNA synthesis (see Table 3.2).

The synthesis of RNA proceeds principally by a similar mechanism. Cells contain a DNA-dependent **RNA polymerase**, which requires GTP, CTP, ATP, and UTP as substrates. Thus RNA is different from DNA because uracil is present in place of thymine. This does not affect pairing with adenine since the structures of uracil and thymine differ by only one methyl group (see Table 3.4). RNA polymerase transcribes only one of the DNA strands and chain elongation proceeds in the 5'→3' direction. Furthermore, the entire DNA genome is not transcribed into RNA sequences; genes which need not be expressed because of the physiological conditions of the cell remain silent. Also, genes which code for a protein required in large amounts are transcribed more frequently than others. Thus the RNA polymerase must receive signals "telling" it which DNA strand to transcribe and at what point transcription is to be initiated and terminated. **Transcription** is, therefore, more complicated than replication.

Likewise, RNA polymerase has a very complicated structure; it consists of six subunits, two α-subunits and one each of β, β', ω, and σ subunits. Thus five different polypeptide chains are present in the enzyme molecule. The so-

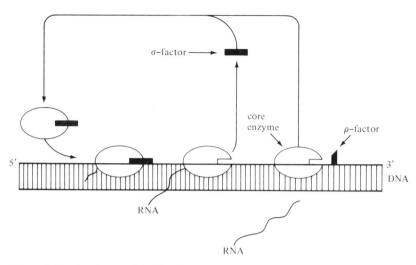

Figure 3.33. Synthesis of RNA. The core enzyme of RNA polymerase combines with the σ-factor. The latter recognizes the start signal at the DNA. During RNA synthesis the σ-factor is released. RNA synthesis is terminated by the p-factor.

called core enzyme $\alpha_2\beta\beta'\omega$ (without σ) exhibits catalytic activity but it is unable to recognize start signals at the DNA. Yet aggregation of the σ-factor with the core enzyme yields the physiologically competent RNA polymerase, and the σ-factor, therefore, causes recognition of the regions at the DNA where the initiation of RNA synthesis is to take place. Recognition of the stop signal for RNA synthesis, at least in some instances, involves a protein accessory factor to trigger RNA polymerase to stop at defined points. Figure 3.33 summarizes the DNA-dependent RNA synthesis.

RNA synthesis requires the same amount of ATP per monomer that is needed for DNA synthesis. From the three classes of RNA formed, **transfer RNA, ribosomal RNA**, and **messenger RNA**, the latter has only a short life (half-life in *E. coli*: 3 min) and is rapidly degraded into nucleoside monophosphates; following their phosphorylation to the corresponding triphosphates, they can be used again for RNA synthesis. When a culture of *E. coli* doubles its cell mass in 30 min it has synthesized its original messenger RNA content approximately 10 times! In terms of ATP requirement, RNA synthesis is about 12 times more expensive for the cells than DNA synthesis (see Table 3.2).

2. Protein synthesis. In replication and transcription, a defined sequence of bases in DNA functions as a template for the synthesis of DNA and RNA molecules, which also have a defined sequence of bases. Both processes can be understood on the basis of specific base pairing. It is obvious that the synthesis of proteins consisting of a certain sequence of amino acids is more difficult to understand, as it is impossible for an amino acid itself to recognize

Figure 3.34. Folded polypeptide chain of an enzyme protein. ‖, Disulfide bridges; ▯, active site.

a certain succession of bases in RNA. Thus the process (called **translation**) of conversion of a sequence of bases into a sequence of amino acids making up a polypeptide chain is very complicated.

All protein functions are determined by their amino acid sequence. Figure 3.34 is a model of an enzyme protein. The proper position of certain amino acids is responsible for correct folding of the polypeptide chain, for the formation of interchain covalent bonds (for instance S-S-bridges), and the function of a certain region of the enzyme molecule as active site. Clearly, the synthesis of biologically active proteins has to proceed with great precision. It is possible that the incorporation of glutamate instead of lysine at a certain position of the polypeptide chain could result in a protein lacking enzyme activity.

In 1961 it was shown by Matthaei and Nierenberg that the formation of polyphenylalanine could be catalyzed in a cell-free system when polyuridylic acid was added as template; other amino acids were not polymerized. Since it was already known by then that units of three nucleotides of RNA (**triplets**) code for one amino acid, it was concluded that the triplet UUU is the code word for phenylalanine. With the help of synthetically prepared ribonucleotides of known base sequences it was then possible to unravel the meaning of all 64 (4^3) possible combinations of triplets. Some code words easy to remember are given in Table 3.6.

How does a base triplet bring about the incorporation of a particular amino acid into a growing polypeptide chain? The principle is indicated in Figure 3.35. Several components are involved: **messenger-RNA**, which functions as the template; **70 S ribosomes** attached to it, which catalyze the formation of peptide bonds and consist of 30 S and 50 S subunits; **transfer-RNAs**, which are linked to the corresponding **amino acids** by specific enzymes—the **aminoacyl-t-RNA synthetases**. In the example shown in Figure 3.35 phenylalanine-specific synthetase links this particular amino acid to the phenylalanine-specific t-RNA. The latter recognizes with its anticodon (AAA) region, by base pairing, the phenylalanine-codon (UUU) of the messenger-RNA. The peptide bond is formed by transferring the polypeptide

Table 3.6. The meaning of some base triplets

DNA	messenger RNA	amino acid
AAA	UUU	phenylalanine
TTT	AAA	lysine
GGG	CCC	proline
TAC	AUG	methionine

chain from the adjacent t-RNA (proline-specific-t-RNA in Figure 3.35) to the amino group of phenylalanine.

The formation of aminoacyl-t-RNA is catalyzed by aminoacyl-t-RNA synthetases according to the following equations:

$$R—\underset{\underset{NH_2}{|}}{CH}—COOH + ATP + enzyme \longrightarrow [R—\underset{\underset{NH_2}{|}}{CH}—CO—AMP]\text{-enzyme} + PP_i$$

$$[R—\underset{\underset{NH_2}{|}}{CH}—CO—AMP]\text{-enzyme} + t\text{-RNA} \longrightarrow R—\underset{\underset{NH_2}{|}}{CH}—CO—t\text{-RNA} + enzyme + AMP$$

The aminoacyl group is attached to the 3'-OH group of the terminal adenosine of t-RNA. *E. coli* contains more than 50 different t-RNAs, so that for each amino acid one to three specific t-RNAs are present. Also present are at least 20 different aminoacyl-t-RNA synthetases, and they are very specific. Each of them is able to pick "its" amino acid out of the 20

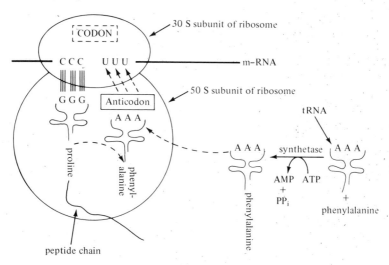

Figure 3.35. Principle of translation.

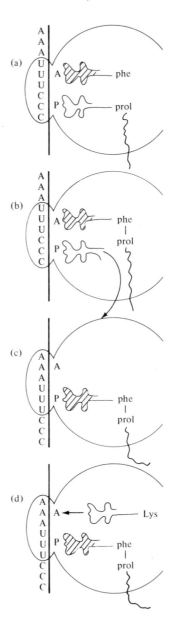

Figure 3.36. Steps involved in the elongation of a peptide by one amino acid. (a) Phe-t-RNA is bound at site A. GTP is required in this step; it is hydrolyzed to GDP and P_i. (b) The peptide bond is formed by transfer of the peptide to phe-t-RNA. Proline-specific t-RNA is released. (c) The translocation reaction. The ribosome moves up one step, so that the peptidyl-t-RNA binds at the P-site. This reaction is also accompanied by GTP hydrolysis. (d) Lys-t-RNA approaches the A-site.

around in the cytoplasm, to form the corresponding adenylate and to transfer the aminoacyl residue to the "correct" t-RNA.

The catalytic machinery for polypeptide synthesis is harbored in ribosomes, which consist of heavy (50 S) and light (30 S) subunits and contain proteins and ribosomal RNA. Peptide bond formation is accomplished by an alternation between two sites of the ribosome, the **aminoacyl (A) site** and the **peptidyl (P) site**. One complete cycle is illustrated in Figure 3.36. From the mode of the transferase reaction (step 2) it is obvious that the polypeptide chain grows from the NH_2-terminal amino acid toward the COOH-terminal one.

The synthesis of all proteins of *E. coli* and other bacteria is initiated by the binding of **N-formylmethionyl-t-RNA** at the peptidyl site of the 30 S subunit, which is attached to the initiation triplet of m-RNA (AUG or GUG). Subsequently the 50 S subunit is bound, the correct aminoacyl-t-RNA approaches the A site of the ribosome, and the first peptide bond is formed. Termination of a polypeptide chain is signaled by special codons of the m-RNA (UAG or UAA). Most proteins do not contain N-formylmethionine as N-terminal amino acid. The formyl group and in many cases the methionine residue are removed from the finished proteins by specific enzymes.

Much energy is required for protein synthesis. The activation of an amino acid is accomplished by the conversion of 1 ATP to $AMP + PP_i$; in one elongation cycle 2 GTP are hydrolyzed to $2 GDP + 2P_i$. Thus the equivalent of 4 ATP is required for the formation of one peptide bond. This makes protein biosynthesis the most expensive process in the cell. *E. coli* growing on glucose has to invest about 50% of its metabolic energy into protein synthesis.

X.   The Requirement for an Anaplerotic Sequence

It has been pointed out that intermediates of the tricarboxylic acid cycle, such as α-oxoglutarate, serve as precursors in the biosynthesis of several cellular constituents. Consequently, the tricarboxylic acid cycle has a dual function: it regenerates reduced pyridine and flavine nucleotides for the respiratory chain and supplies starting material for biosyntheses. This latter function creates a problem, as the principle of the cycle is that a C_4-dicarboxylic acid condenses with a C_2 compound to yield a C_6 compound; the latter is oxidized in such a way that the C_4-dicarboxylic acid is regenerated. If an intermediate such as α-oxoglutarate is drained off for biosynthetic purposes, the pool of C_4-dicarboxylic acids must collapse. Clearly, additional reactions have to be present, which form C_4-dicarboxylic acids directly from intermediates of glucose breakdown. According to H. L. Kornberg, reactions of this type are called **anaplerotic sequences** (replenishing reactions). The situation in *E. coli* growing with glucose is illustrated in

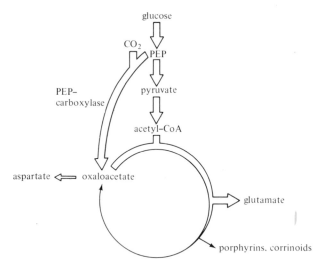

Figure 3.37. The function of PEP carboxylase as anaplerotic enzyme during growth of *E. coli* on glucose.

Figure 3.37. Cycle intermediates are used for the synthesis of glutamate, porphyrins, and aspartate; the oxaloacetate needed is supplied by the PEP-carboxylase reaction:

$$PEP + CO_2 \xrightarrow{\text{PEP carboxylase}} oxaloacetate + P_i$$

Mutants of *E. coli* lacking this enzyme only grow on glucose when the medium is supplemented with utilizable intermediates of the tricarboxylic acid cycle or precursors thereof. As will be discussed later other organisms employ pyruvate carboxylase as anaplerotic enzyme during growth on carbohydrates, and during growth on substrates like acetate or pyruvate different anaplerotic sequences are effective.

XI. Summary

1. *Escherichia coli* is able to synthesize all its cellular constituents from glucose and minerals. The strategy of biosynthesis is that relatively few intermediates of the energy metabolism (glucose-6-phosphate, phosphoenolpyruvate, α-oxoglutarate, etc.) are used to synthesize monomers, which subsequently are polymerized to give macromolecules; the latter comprise approximately 97% of the dry weight of the cells.

2. At low concentrations of ammonia, it is assimilated by glutamine synthetase. Glutamate is formed by the transfer of the amido group to α-oxoglutarate (glutamate synthase reaction). At high concentrations of

ammonia, glutamate is formed from α-oxoglutarate also by glutamate dehydrogenase.

3. The principal sulfur source is sulfate. It is reduced to H_2S via APS and PAPS. H_2S reacts with O-acetylserine to yield cysteine.

4. Twenty amino acids are required for protein synthesis. Alanine and aspartate are synthesized from pyruvate and oxaloacetate by transamination with glutamate as NH_2-donor. Asparagine is formed in a reaction analogous to the glutamine synthetase reaction. Reduction of aspartate yields aspartic semialdehyde—the precursor of lysine, threonine, and methionine. Deamination of threonine gives α-oxobutyrate, which by the successive action of four enzymes is converted into isoleucine. The same four enzymes convert pyruvate into valine; an intermediate in valine synthesis serves also as precursor in leucine formation. Serine, glycine, and cysteine are formed from 3-phosphoglycerate, and proline and arginine from glutamate. The three aromatic amino acids, tyrosine, phenylalanine, and tryptophan, are synthesized from erythrose-4-phosphate and PEP; shikimate and chorismate are common intermediates. Histidine is formed from 5-phosphoribosyl-1-pyrophosphate and ATP in a complex series of reactions by the action of nine enzymes.

5. Pentose phosphates and $NADPH_2$ are formed in the oxidative pentose phosphate cycle. Reactions yielding $NADPH_2$ are the glucose-6-phosphate and 6-phosphogluconate dehydrogenase reactions. The first pentose phosphate formed is ribulose-5-phosphate; it can be isomerized to yield ribose-5-phosphate. Pentose phosphates can be converted back into hexose phosphates by the transketolase and transaldolase reactions.

6. Aspartate, carbamyl phosphate, and 5-phosphoribosyl-1-pyrophosphate are the precursors of the pyrimidine nucleotides. First orotic acid is formed; it is converted to orotidine monophosphate and finally to UTP and CTP. An imidazole ribonucleotide is an intermediate in purine nucleotide synthesis. Subsequent carboxylation, formylation, and amination reactions yield inosinic acid (IMP), which is the ultimate precursor of ATP and GTP. Reduction of ribonucleotides to deoxyribonucleotides takes place at the diphosphate level.

7. *Escherichia coli* contains phospholipids and glycolipids as constituents of membrane systems. Neutral fats are not stored. Long-chain fatty acids are synthesized by soluble enzyme systems from acetyl-ACP (acyl carrier protein) and malonyl-ACP. From the finished acyl-ACP the acyl residue is transferred to glycerophosphate to yield phosphatidic acids. Esterification of the latter with alcohols yields the various phospholipids.

8. UDP-N-acetylglucosamine and its 3-lactyl derivative, UDP-N-acetylmuramic acid, are precursors of the peptidoglycan. They are synthesized from fructose-6-phosphate. UDP-N-acetylmuramic acid is further converted to UDP-N-acetylmuramyl-pentapeptide. The pentapeptide chain of *E. coli* consists of L-alanine, D-glutamate, *m*-diaminopimelate, D-alanine, and D-alanine.

9. Membranes like the cytoplasmic membrane consist of a double layer of phospholipid molecules; they contain up to 60% protein. Membranes have specific functions in transport processes.

10. Periodic macromolecules such as glycogen, peptidoglycan, and the constituents of the outer membrane layer are synthesized from their building blocks by specific enzymes. Glycogen is formed by transfer of glucose from ADP-glucose to a primer oligosaccharide. The two building blocks of peptidoglycan are first transferred to the membrane lipid carrier undecaprenyl phosphate (bactoprenol) and then transported to the elongation site of the peptidoglycan. Polysaccharide bonds are formed and cross-linkage between the polysaccharide chains is accomplished by peptide bonds between the *m*-diaminopimelate residue of one chain and the D-alanine residue of another chain. The outer membrane layer consists of lipid A, the core oligosaccharide, and repeating oligosaccharide chains.

11. The synthesis of informational macromolecules proceeds at templates. DNA serves as template for DNA and RNA synthesis and messenger-RNA serves as template for protein synthesis. Information is stored in DNA in the form of defined base sequences; these sequences are recognized by base pairing between adenine and thymine (uracil in the case of RNA) and between guanine and cytosine. Substrates in DNA and RNA synthesis are the ribonucleoside and deoxyribonucleoside triphosphates; pyrophosphate is formed in the polymerization reactions.

In the translation of base sequences into amino acid sequences base triplets of messenger-RNA serve as code words for the amino acids. The base triplets are recognized by base pairing with the anticodon region of transfer-RNA. For instance, the triplet UUU is recognized by the AAA-anticodon of phenylalanine-specific transfer-RNA. Thus phenylalanine is incorporated into a growing peptide chain when UUU appears at the messenger-RNA. Protein synthesis proceeds at the ribosomes, which consist of 50 S and 30 S subunits. Protein synthesis is initiated by the binding of N-formylmethionyl-t-RNA at the 30 S ribosomal subunit, which is attached to the initiation triplet of m-RNA. Subsequently the 50 S subunit is bound and the formation of peptide bonds begins. The equivalence of 4 ATP is required for the formation of one peptide bond.

12. When *E. coli* grows on glucose the PEP-carboxylase reaction serves as an anaplerotic reaction.

Chapter 4

Aerobic Growth of *Escherichia coli* on Substrates Other Than Glucose

Escherichia coli is able to grow on carbohydrates other than glucose and on a number of organic acids and amino acids. The additional enzyme systems required for the utilization of some of these substrates will be discussed in this chapter.

I.   Fructose and Lactose as Substrates

Fructose is an excellent substrate for the growth of many bacteria. As is the case with glucose, *E. coli* employs a PEP-phosphotransferase system to bring fructose into the cell. However, the product appearing inside the cell is fructose-1-phosphate and not the 6-phosphate:

$$PEP + HPr \underset{}{\overset{\text{enzyme I}}{\rightleftharpoons}} \text{phospho-HPr} + \text{pyruvate}$$

$$\text{phospho-HPr} + \text{fructose} \xrightarrow{\text{enzyme II}} \text{fructose-1-phosphate} + HPr$$

Consequently, in addition to the fructose-specific enzyme II of the transport system, a second enzyme is required for the utilization of fructose: **1-phosphofructokinase**. The synthesis of this enzyme is specifically induced by fructose. Figure 4.1 illustrates the difference in the first steps of glucose and fructose metabolism. From this it becomes clear as to why Kornberg and associates were able to isolate mutants that grew on glucose but not on fructose. These mutants lacked 1-phosphofructokinase.

Glucose-grown cells of *E. coli* are not able to utilize **lactose** immediately. To do so several additional enzymes are required, and their formation is induced when lactose becomes available to cells.

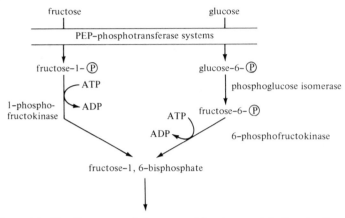

Figure 4.1. The first steps of glucose and fructose metabolism in *E. coli*.

Three enzymes are synthesized following the addition of lactose to the growth medium: **lactose permease, β-galactosidase**, and **β-galactoside acetyltransferase**. These enzymes are of great importance, not only because they enable *E. coli* to grow on lactose but also because the investigation of their coordinate formation has led to the **operon model** of Jacob and Monod and to an understanding of the regulation of enzyme synthesis. This will be discussed in Chapter 7.

Lactose permease catalyzes an energy-dependent transport of lactose into the cell and β-galactosidase hydrolyses lactose to yield glucose and galactose (Figure 4.2). The biological function of the β-galactoside acetyltransferase, which acetylates lactose and other β-galactosides with acetyl-CoA, is

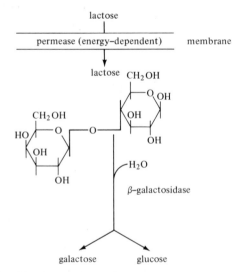

Figure 4.2. Conversion of lactose to galactose and glucose.

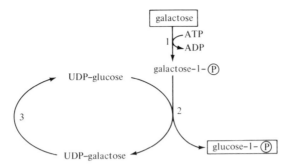

Figure 4.3. Conversion of galactose into glucose-1-phosphate. 1, galactokinase; 2, glucose:galactose-1-phosphate uridylyltransferase; 3, uridine-diphosphate glucose epimerase.

probably that of a detoxification enzyme: nonmetabolizable structural analogues of lactose are acetylated and excreted.

The glucose formed from lactose is phosphorylated to glucose-6-phosphate by hexokinase. It should be recalled that hexokinase is not required by *E. coli* growing on glucose or fructose (the PEP-phosphotransferase systems yield hexose monophosphate directly); however, hexokinase is employed for phosphorylation of glucose or fructose formed inside the cells by oligosaccharide hydrolysis.

The galactose formed from lactose induces the synthesis of three enzymes involved in its further metabolism: **galactokinase** (1), **glucose:galactose-1-phosphate uridylyltransferase** (2), and **UDP-glucose epimerase** (3). Together they catalyze the following reactions: Galactose is phosphorylated at carbon atom 1; UDP-glucose reacts with galactose-1-phosphate to yield glucose-1-phosphate and UDP-galactose; the latter is epimerized to UDP-glucose (Figure 4.3).

With the induction of all these enzymes it then becomes possible to feed lactose into the Embden–Meyerhof and subsequent pathways.

II. Acetate, Pyruvate, and Malate as Substrates

When *E. coli* is transferred from a glucose or fructose medium to a medium containing acetate as carbon and energy source, this change is rather drastic as all cellular constituents must then by synthesized starting from C_2-units. Obviously this will require a number of new enzyme systems, a description of which now follows. The formation of $NADH_2$ to be used for ATP production in the respiratory chain is very simple. Acetate is actively transported into the cell and subsequently converted into acetyl-CoA:

$$\text{acetate} + \text{ATP} \xrightarrow{\text{acetate kinase}} \text{acetyl phosphate} + \text{ADP}$$

$$\text{acetyl phosphate} + \text{CoA} \xrightarrow{\text{phosphotransacetylase}} \text{acetyl-CoA} + P_i$$

Acetyl-CoA then is fed into the tricarboxylic acid cycle, thus yielding $NADH_2$.

The problem, however, is how intermediates of the cycle which serve as starting materials in biosyntheses are regenerated. Clearly, the anaplerotic sequence yielding oxaloacetate from PEP, which is important during growth on carbohydrates, cannot be functional; there is no direct way to synthesize PEP from acetate. Two other anaplerotic enzymes are employed instead: **isocitrate lyase** and **malate synthase**. The first of these catalyzes the cleavage of isocitrate to succinate and glyoxylate:

$$
\begin{array}{ccc}
\text{CH}_2\text{—COOH} & & \text{CH}_2\text{—COOH} \\
| & \xrightarrow{\text{isocitrate lyase}} & | \\
\text{HC—COOH} & \rightleftharpoons & \text{CH}_2\text{—COOH} \\
| & & + \\
\text{HO—CH—COOH} & & \text{O}=\text{CH—COOH}
\end{array}
$$

Malate synthase catalyzes the condensation of acetyl-CoA with glyoxylate to yield malate—a reaction analogous to the citrate synthase reaction.

$$
\begin{array}{ccc}
\text{CH}_3\text{—CO—CoA} & & \text{CH}_2\text{—COOH} \\
+ & \xrightarrow[\text{malate synthase}]{+\text{H}_2\text{O}} & | \quad\quad + \text{CoA} \\
\text{O}=\text{CH—COOH} & & \text{HO—CH—COOH}
\end{array}
$$

Together with enzymes of the tricarboxylic acid cycle these reactions form the so-called **glyoxylate cycle**. The significance of this cycle is illustrated in Figure 4.4. Reactions of the TCA cycle plus isocitrate lyase catalyze the oxidation of acetyl-CoA to glyoxylate. The latter can then condense with another molecule of acetyl-CoA to yield malate. Thus, extra oxaloacetate becomes available (actually formed from two acetyl-CoA), which can be used to form glutamate (via citrate and α-oxoglutarate) and other compounds derived from cycle intermediates. Moreover, oxaloacetate serves as starting material for all the carbohydrates required for polymer synthesis. The ultimate precursor of **gluconeogenesis** is phospho-enolpyruvate, and it is formed from oxaloacetate and ATP by PEP carboxykinase:

$$\text{oxaloacetate} + \text{ATP} \underset{\xrightarrow{\text{PEP carboxykinase}}}{\rightleftharpoons} \text{phospho-enolpyruvate} + \text{ADP} + \text{CO}_2$$

All the reactions of the Embden–Meyerhof pathway between PEP and fructose-1,6-bisphosphate are reversible so that under the appropriate conditions the latter can be synthesized from PEP. Fructose-1,6-bisphosphate is taken out of equilibrium by the action of a specific phosphatase, which irreversibly hydrolyzes the bisphosphate to fructose-6-phosphate.

The glyoxylate cycle reactions serve as an anaplerotic sequence when *E. coli* grows on acetate or on compounds that are degraded via acetyl-CoA.

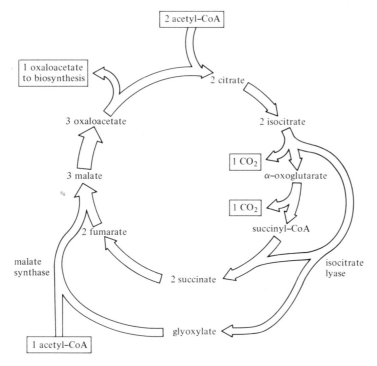

Figure 4.4. The function of the tricarboxylic acid cycle and the glyoxylate cycle reactions during growth of *E. coli* on acetate. For illustration 2 oxaloacetates are allowed to react with 2 acetyl-CoA to yield 2 citrates. These are converted into 2 isocitrates, one of which is oxidized through the tricarboxylic acid cycle and regenerates 1 oxaloacetate. The second is cleaved into succinate and glyoxylate; oxidation of succinate yields a second molecule of oxaloacetate. Glyoxylate condenses with acetyl-CoA to form malate, from which a third molecule of oxaloacetate can be formed. Thus one oxaloacetate is available for biosynthetic purposes.

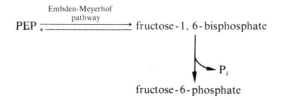

Surprisingly, *E. coli* employs another anaplerotic sequence when growing on pyruvate. This substrate is oxidized to acetyl-CoA by the pyruvate dehydrogenase complex, which is subsequently fed into the TCA cycle. In the synthesis of PEP a novel enzyme is involved: **PEP synthetase**. In this connection it should be recalled that in the Embden–Meyerhof pathway pyruvate kinase catalyzes the conversion of PEP into pyruvate, a reaction accompanied by a decrease of the standard free energy at pH 7 of 5.7 kcal/ mol (23.8kJ/mol)

$$PEP + ADP \xrightarrow{\text{pyruvate kinase}} \text{pyruvate} + ATP \qquad \Delta G_0' = -5.7 \text{ kcal/mol}$$
$$(-23.8kJ/mol)$$

Thus, this reaction is irreversible and cannot provide the PEP for gluconeogenesis. In the PEP synthetase reaction a negative free energy change is achieved by the hydrolysis of one of the energy-rich pyrophosphate bonds of ATP.

$$\text{pyruvate} + ATP \xrightarrow[\text{synthetase}]{\text{PEP}} PEP + AMP + P_i \qquad \Delta G_0' = -2 \text{ kcal/mol}$$
$$(-8.4kJ/mol)$$

A pyrophosphorylated enzyme is an intermediate in this reaction:

$$\text{enzyme} + ATP \rightleftharpoons \text{enzyme-PP} + AMP$$

$$\text{enzyme-PP} + H_2O \rightleftharpoons \text{enzyme-P} + P_i$$

$$\text{enzyme-P} + \text{pyruvate} \rightleftharpoons PEP + \text{enzyme}$$

The oxaloacetate required to replenish the pool of TCA cycle intermediates is formed, as it is during growth on glucose and other carbohydrates, from PEP by PEP carboxylase. The importance of PEP synthetase during growth of *E. coli* on pyruvate or lactate is well documented. Mutants were isolated, which grew on glucose and acetate but not on pyruvate and lactate; they lacked PEP synthetase.

C_4-dicarboxylic acids are good substrates for aerobic growth of *E. coli*. L-Malate, for instance, is taken up and oxidized to oxaloacetate by malate dehydrogenase. Oxaloacetate thus formed can serve as substrate in anabolic reactions and in the tricarboxylic acid cycle. However, it is rather obvious that an enzyme system is required that converts malate into pyruvate; otherwise acetyl-CoA and the various metabolites derived from pyruvate could not be formed. The synthesis of pyruvate is accomplished by malic enzyme:

$$\text{L-malate} + NAD \xrightleftharpoons{\text{malic enzyme}} \text{pyruvate} + NADH_2 + CO_2$$

In addition to the NAD-specific enzyme, an NADP-specific malic enzyme is present in *E. coli* cells growing with malate; it provides $NADPH_2$ for biosyntheses. PEP is formed from oxaloacetate by PEP carboxykinase and from pyruvate by PEP synthetase.

The fraction of ATP which has to be invested in the biosynthesis of monomers during growth depends very much on the nature of the growth substrate; it is small when glucose is the substrate because the formation of the precursors in monomer synthesis (PEP, pyruvate, acetyl-CoA) is connected with a gain of ATP. On the other hand, the formation of the same compounds from acetate requires ATP. Stouthamer calculated that 34.8

mmol of ATP are required for the formation of 1 g of cells from glucose and 99.5 mmol of ATP for the formation of 1 g of cells from acetate.

III.   Summary

1. The change of the growth substrate necessitates changes in the enzyme equipment of the cells: (1) synthesis of the appropriate catabolic enzymes and the substrate-specific transport systems; (2) formation of the appropriate anaplerotic sequences.

2. Fructose transport by the PEP-phosphotransferase system yields fructose-1-phosphate, which is phosphorylated by 1-phosphofructokinase to fructose-1,6-bisphosphate. Lactose is transported by lactose permease into the cells and subsequently cleaved by β-galactosidase to glucose and galactose.

3. During growth on acetate *E. çoli* requires the glyoxylate cycle as anaplerotic sequence. It consists of enzymes of the tricarboxylic acid cycle plus isocitrate lyase and malate synthase and accomplishes the synthesis of C_4-dicarboxylic acids from acetate. In addition, PEP carboxykinase is required for PEP formation in gluconeogenesis.

4. During growth of *E. coli* on pyruvate, PEP synthetase and PEP carboxylase serve as anaplerotic enzymes; during growth on L-malate, malic enzyme is required for the formation of pyruvate.

5. The fraction of ATP which has to be invested in the biosynthesis of monomers during growth is dependent on the nature of the growth substrate.

Chapter 5

Metabolic Diversity of Aerobic Heterotrophs

In the preceding chapters we have used *E.coli* as a model organism in order to become acquainted with the principal reactions participating in the ATP production during aerobic growth and in the biosynthesis of cellular material. Furthermore, we have discussed the changes in the enzymatic machinery of the cell that must occur when glucose is replaced by other substrates.

This relatively simple picture becomes much more complicated when we consider other bacteria. Many of them resemble *E. coli* in their nutritive requirements but use modified catabolic pathways and alternate anaplerotic sequences or form different storage materials and different cell wall components. Some of this metabolic diversity among aerobic heterotrophs will be discussed in this chapter.

I. The Different Mechanisms for the Uptake of Substrates

E. coli cells take up glucose and fructose by phosphotransferase systems and lactose, in contrast, by active transport. When comparing the mode in which substances from the environment cross the cell membrane into the cytoplasm four different mechanisms can be distinguished: passive diffusion, facilitated diffusion, active transport, and group translocation.

A. Passive diffusion

The transported substance does not specifically interact with components of the cell membrane. It crosses the membrane until equilibrium is reached between the concentration inside and that outside. In that the concentration of most of the metabolites in nature is higher inside the cell than outside, it is clear that transport by passive diffusion must be restricted to a small group of substances, i.e., gases such as oxygen, water, and some ions. In *E. coli*, sodium ions are taken up by passive diffusion, while potassium and magnesium ions are actively transported into the cell. However, it also should be mentioned that many other bacteria employ active transport systems for sodium ions.

B.   Facilitated diffusion

This process is similar to passive diffusion in that neither one requires metabolic energy, and both are freely reversible such that equal concentrations of the substance are found inside and outside the cell. However, unlike passive diffusion, facilitated diffusion involves transport of a substance via a **specific membrane carrier**; the substance is bound to the carrier on the outside and released on the inside of the cell. Transport is, therefore, substrate-specific. If, for instance, a membrane contains only a glucose-specific carrier, other carbohydrates will not be transported into the cell. Erythrocytes and yeast cells take up sugars by facilitated diffusion, while in aerobic bacteria, this transport mechanism does not seem to be very important. However, in anaerobes the process appears to be involved in the uptake of certain compounds and in the excretion of fermentation products.

C.   Active transport

Figure 5.1 illustrates the effectiveness of active transport as compared to the diffusion processes.

Clearly, active transport allows substrate saturation of the cell's enzymatic

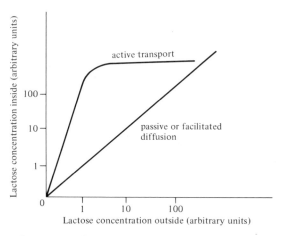

Figure 5.1. Internal concentration of lactose in *E. coli* as dependent on the lactose concentration in the medium and the transport process involved.

machinery at much lower concentrations of the substrate in the medium than do the diffusion processes. A several hundredfold concentration of compounds has been observed; thus active transport enables the cells to live nicely in environments containing substrates in low concentrations, a situation common in nature. An uptake process is defined as active transport according to the following parameters:

1. It is substrate-specific; a carrier-substrate complex is formed on the outside surface of the membrane.
2. It requires metabolic energy; the carrier has a high affinity for the substrate when facing the outside surface of the membrane and a low affinity for it when facing the inside surface. This modification of the carrier requires energy.
3. It allows transport of its substrate against a concentration gradient; this results from the change in substrate affinity of the carrier when turning from outside to inside.
4. It releases the substrate unmodified into the cytoplasm (different from group translocation).

Active transport is schematically illustrated in Figure 5.2.

What is the energy source of active transport? In most systems it apparently is the **protonmotive force**, which can be generated either by electron transport or by ATP hydrolysis as catalyzed by the membrane-bound ATPase BF_1 [Figure 5.3(a) and (b)]. The protonmotive force is taken advantage of by carriers having binding sites for protons and a particular substrate, and both (protons and substrate molecules) are transported into

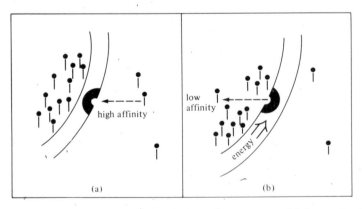

Figure 5.2. Uptake of a substrate molecule by active transport. (a) the carrier faces the outside surface of the membrane and has a high affinity for substrate binding; (b) at the expense of metabolic energy the carrier-substrate complex is modified such that it faces the inside surface of the membrane. At the inside of the membrane the affinity of the carrier for the substrate is sufficiently lowered so that the complex dissociates even at high substrate concentrations.

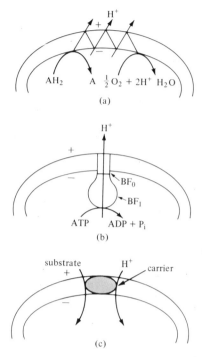

Figure 5.3. The protonmotive force as the energy source for active transport. (a) generation of the protonmotive force by respiration; (b) generation of the protonmotive force by ATP hydrolysis; (c) active transport by a carrier with binding sites for a particular substrate and for protons.

the cells [Figure 5.3(c)]. The protonated form of the carrier must have a high affinity for the substrate and the deprotonated form a low affinity.

That electron transport or ATP can provide the energy for active transport is important when we consider that not only aerobes and phototrophs but also anaerobes require such transport systems. Many of the latter organisms gain ATP only by substrate-level phosphorylation and do not contain electron transport chains. Figure 5.4 summarizes the energy sources for active transport across an energized membrane.

There is a second active transport system present in *E. coli* and other microorganisms (particularly in Gram-negative bacteria). It is involved in the transport of amino acids, peptides, some sugars, and organic acids and consists of a substrate-specific **binding protein** and an ATP-requiring transport unit. Binding proteins differ from the carriers of the electron transport coupled system in that they are readily released from the periplasmic space by Heppel's osmotic shock procedure. According to this method, bacteria suspended in a hypertonic solution (a 20% solution of glucose with buffer and EDTA) are suddenly shifted to a cold solution of low ionic strength

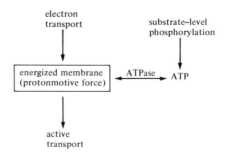

Figure 5.4. Electron transport and ATP as energy sources for active transport.

(water or a 5-mM MgCl$_2$ solution); binding proteins then appear in solution. Compounds such as glutamine, proline, galactose, and maltose are actively transported into *E. coli* cells via binding proteins.

D.   Group translocation

This process differs from active transport in that the substrate appears inside the cell in a **chemically modified form**—usually as phosphate ester. Sugars are transported in many microorganisms by group translocation. The driving force of transport in this case is that the sugar is trapped within the membrane by reaction with a phosphorylated enzyme and that the phosphate ester formed is released into the cytoplasm. The phosphorylated enzyme is

Table 5.1. PEP phosphotransferase and active transport systems for glucose in bacteria[a]

organism	phospho-transferase	active transport
Arthrobacter species	+	−
Azotobacter vinelandii	−	+
Bacillus megaterium	+	−
B. subtilis	+	−
Brucella abortus	+	−
Clostridium butyricum	+	−
C. pasteurianum	+	−
Enterobacteria	+	−
Micrococcus luteus	−	+
Mycobacterium smegmatis	−	+
Pseudomonas aeruginosa	−	+
Staphylococcus aureus	+	−

[a]Most data are from A. H. Romano, S. J. Eberhard, S. L. Dingle, and T. D. McDowell, *J. Bacteriol.* **104**, 808–813 (1970).

generated using PEP as the principal source of phosphate-bound energy. The mechanism of the **PEP phosphotransferase** systems has been outlined in Chapter 2, Section I (glucose) and Chapter 4, Section I (fructose). In *E. coli*, glucose, fructose, mannose, and mannitol are transported by group translocation and in *Staphylococcus aureus* lactose and other disaccharides also are thus transported.

A phosphotransferase system is, however, not present in all bacteria that are able to grow on glucose, as can be seen in Table 5.1. It appears that this system occurs predominantly in facultatively and strictly anaerobic bacteria. Aerobes such as *Azotobacter vinelandii* and *Pseudomonas aeruginosa* employ an active transport system for the uptake of glucose. There are, of course, exceptions: *Arthrobacter* species, *Bacillus megaterium*, and *B. subtilis*, which are aerobes, transport glucose by a phosphotransferase system into the cell. For anaerobes, the phosphotransferase system is of great importance as it helps to save ATP.

There is also evidence that purine and pyrimidine bases are transported by group translocation. The uptake of adenine by *Escherichia coli* and *Salmonella typhimurium* is mediated by an enzyme system requiring 5-phosphoribosyl-1-pyrophosphate; AMP appears inside the cell:

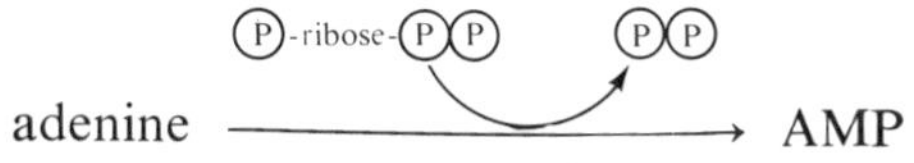

$$\text{(P)-ribose-(P)(P)} \qquad \text{(P)(P)}$$

$$\text{adenine} \xrightarrow{\hspace{4cm}} \text{AMP}$$

II.   The Entner—Doudoroff Pathway

Besides the Embden–Meyerhof pathway, there is a second important pathway used for carbohydrate breakdown, which is widely distributed among bacteria. It was first discovered by Entner and Doudoroff in *Pseudomonas saccharophila*. As in the pentose phosphate cycle, glucose-6-phosphate is first dehydrogenated to yield 6-phosphogluconate (Figure 5.5). This is converted by a dehydratase and an aldolase reaction into 1 molecule of 3-phosphoglyceraldehyde and 1 molecule of pyruvate. The 3-phosphoglycer-

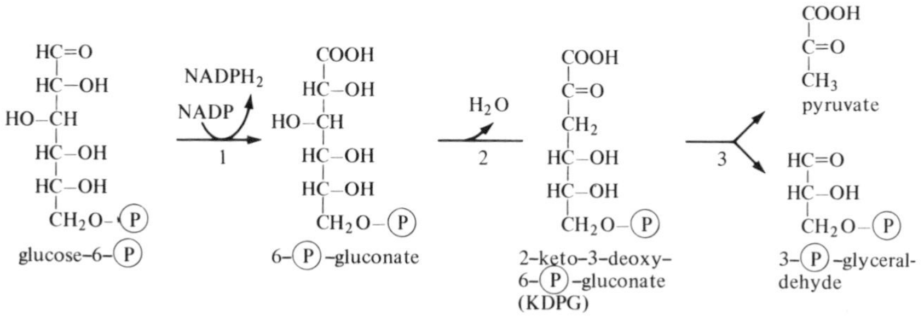

Figure 5.5. Reactions of the Entner–Doudoroff pathway. 1, glucose-6-phosphate dehydrogenase; 2, 6-phosphogluconate dehydratase; 3, KDPG aldolase.

aldehyde can then be oxidized to pyruvate by the enzymes of the Embden–Meyerhof pathway.

What are the differences between the Embden–Meyerhof (EM) and the Entner–Doudoroff (ED) pathways and how is it possible to differentiate between them? The key enzymes of the EM pathway—meaning the enzymes common only to this pathway—are 1- and 6-phosphofructokinase, respectively. All the other enzymes of the EM pathway may be part of other metabolic sequences as well (FDP aldolase, for instance, in gluconeogenesis or in the pentose phosphate cycle). The key enzymes of the ED pathway are **6-phosphogluconate dehydratase** and **2-keto-3-deoxy-6-phosphogluconate (KDPG) aldolase**. Thus, these pathways can be distinguished in the following way: cells of a particular bacterium are grown on glucose, cell extracts are prepared, and the key enzymes of the two pathways are assayed. If high levels of the dehydratase and KDPG aldolase are present and phosphofructokinases are not detectable, it is likely that the ED pathway is involved in glucose degradation.

A second method, **radiorespirometry**, might also be used to determine whether the EM or ED pathway exists in a particular organism. The basis for this is illustrated in Figure 5.6. It is obvious that both pathways yield "different" pyruvate from the first 3 carbon atoms of glucose. In one case (EM) the carboxyl group originates from carbon 3 of glucose, in the other (ED) from carbon 1 of glucose. During aerobic oxidation of glucose the first CO_2 appearing comes from the oxidative decarboxylation of pyruvate; it thus originates from glucose carbon atoms 3 and 4 (EM) or 1 and 4 (ED). This difference is detectable by radiorespirometric methods. For instance, glucose labeled with ^{14}C in C_1, C_3, or C_4 is added to cell suspensions of *Bacillus subtilis* and *Xanthomonas phaseoli*. The CO_2 evolved is collected

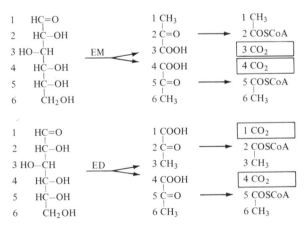

Figure 5.6. Origin of the CO_2-carbon formed in the pyruvate dehydrogenase reaction after degradation of glucose by either the Embden–Meyerhof (EM) or the Entner–Doudoroff (ED) pathway.

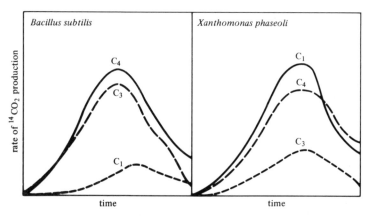

Figure 5.7. Evolution of radioactive CO_2 from glucose labeled in positions 1, 3, or 4 by cell suspensions of *B. subtilis* and *X. phaseoli*. [Data from V. H. Cheldelin, *Metabolic Pathways in Microorganisms*. John Wiley and Sons, Inc., New York, pp. 30—63 (1961) and A. C. Zagallo and C. H. Wang, *J. Bacteriol.* **93**, 970—975 (1967).]

every hour and its radioactivity is determined. It is evident from Figure 5.7 that *B. subtilis* releases radioactivity faster from C_4 and C_3 of glucose and *X. phaseoli* from C_1 and C_4; thus, the first bacterium employs the Embden–Meyerhof and the second the Entner–Doudoroff pathway.

Which of these two pathways is mainly involved in the breakdown of hexoses by various bacteria is summarized in Table 5.2. As can be seen, the Entner–Doudoroff pathway is fairly widespread, particularly among Gram-negative bacteria, while it is not very common in anaerobes. In view of the fact that anaerobes do not have a respiratory chain and gain their ATP usually by substrate-level phosphorylation, the EM pathway is more econ-

Table 5.2. Pathways involved in sugar degradation by bacteria

microorganism	Embden–Meyerhof	Entner–Doudoroff
Arthrobacter species	+	−
Azotobacter chroococcum	+	−
Alcaligenes eutrophus	−	+
Bacillus subtilis, B. cereus + other species	+	−
Escherichia coli + other enterobacteria	+	−
Pseudomonas saccharophila, *P. fluorescens* + other species	−	+
Rhizobium japonicum + other species	−	+
Thiobacillus intermedius, *Th. ferrooxidans*	−	+
Xanthomonas phaseoli + other species	−	+

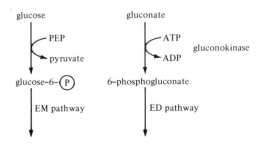

Figure 5.8. Degradation of glucose and gluconate by *E. coli.*

omical for them; the EM pathway yields 2 ATP in the conversion of 1 glucose into 2 pyruvates; the ED pathway yields only 1 ATP per 2 pyruvates.

The ED pathway is also important when compounds such as gluconate, mannonate, or hexuronates serve as substrates. If *E. coli* is transferred from a glucose medium to a gluconate medium, the following enzymes are formed: gluconokinase, which phosphorylates gluconate in position 6, and the two key enzymes of the Entner–Doudoroff pathway. Thus, in *E. coli* glucose and other hexoses are degraded via the EM pathway, but gluconate and related compounds via the ED pathway (Figure 5.8).

Clearly, the EM pathway is not very useful for gluconate degradation. A reaction sequence leading directly from 6-phosphogluconate to glucose-6-phosphate is not known, and many bacteria utilize gluconate, mannonate, glucuronate, and related compounds via the ED pathway.

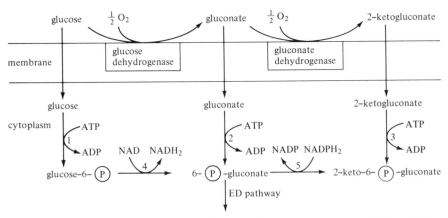

Figure 5.9. Extracellular oxidation of glucose by pseudomonads and intracellular conversion of glucose, gluconate, and 2-ketogluconate into 6-phosphogluconate. 1, hexokinase; 2, gluconate kinase; 3, 2-ketogluconate kinase; 4, glucose-6-phosphate dehydrogenase; 5, 2-keto-6-phosphogluconate reductase. [P. H. Whiting, M. Midgley, and E. A. Dawes, *J. Gen. Microbiol.* **92**, 304–310 (1976).]

It should be mentioned in this connection that a number of pseudomonads also degrade glucose via gluconate. *P. aeruginosa*, *P. fluorescens*, and *P. putida* contain a **glucose dehydrogenase,** which converts glucose to gluconate. The latter can be alternatively phosphorylated to 6-phosphogluconate or further oxidized to 2-ketogluconate, which is subsequently phosphorylated and reduced to yield also 6-phosphogluconate (Figure 5.9). Glucose and gluconate dehydrogenases are membrane-bound. These enzymes allow the extracellular conversion of glucose into compounds that are less utilizable by other microbes than glucose. Thus, these enzymes benefit the pseudomonads, which can transport and degrade glucose, gluconate, and 2-ketogluconate. Fructose is metabolized by these organisms via the conventional Entner–Doudoroff route.

III. Sugar Degradation via the Pentose Phosphate Cycle

As has been outlined in Chapter 3, *E. coli* oxidizes part of its substrate glucose via 6-phosphogluconate to pentose phosphates in order to fulfill the requirements of the growing cell for $NADPH_2$ and precursors of nucleotide biosynthesis. The majority of aerobic microorganisms do the same, as it can be estimated that about 20% of the glucose being degraded flows into this pathway. It also has been pointed out that pentose phosphates can be converted by a series of reversible transaldolase and transketolase reactions into fructose-6-phosphate and 3-phosphoglyceraldehyde. A combination of these reactions results in a cycle that can bring about the degradation of sugars (Figure 5.10). Glucose can be oxidized to 3 CO_2 and glyceraldehyde-3-phosphate, and the latter can be channeled into the tricarboxylic acid cycle via pyruvate. Thus, glucose can be oxidized without participation of either the EM or the ED pathway. This oxidative pentose phosphate cycle is found in *Thiobacillus novellus* and *Brucella abortus*, which lack the key enzymes of the EM and ED pathways but do grow on glucose.

The oxidative pentose phosphate cycle also can allow aerobic growth on carbohydrates without participation of the tricarboxylic acid cycle. If glyceraldehyde-3-phosphate flows back into the hexose-monophosphate pool (via fructose-1,6-bisphosphate), the total oxidation of glucose via this cycle is possible. Alternatively, the C_3 unit can be oxidized to acetate, which is then excreted. This is the situation in some acetic acid bacteria that lack a complete tricarboxylic acid cycle.

Finally it should be mentioned that the nonoxidative reactions of the pentose phosphate cycle participate in the **breakdown of pentoses** in many bacteria. A widespread pathway is: phosphorylation of the pentose in position 5, isomerization and epimerization to the intermediates of the cycle, conversion to fructose-6-phosphate and glyceraldehyde-3-phosphate, degradation by the Embden–Meyerhof pathway.

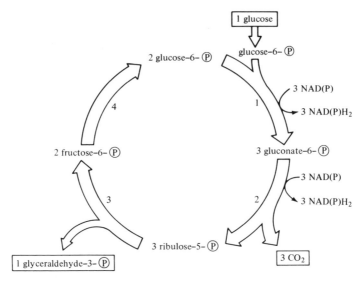

Figure 5.10. Oxidative pentose phosphate cycle. Glucose is oxidized to 3 CO_2 and glyceraldehyde-3-phosphate. 1, glucose-6-phosphate dehydrogenase; 2, gluconate-6-phosphate dehydrogenase; 3, transaldolase and transketolase reactions; 4, phosphoglucose isomerase. NAD(P) means that the dehydrogenase uses either NAD or NADP as coenzymes.

IV. The Methylglyoxal Bypass

In 1923 Harden considered methylglyoxal as an intermediate in glucose catabolism. Following the identification of phosphorylated compounds as intermediates, the accumulation of methylglyoxal observed under certain conditions was considered to result from nonenzymatic reactions. In recent years, however, Cooper and associates discovered a methylglyoxal bypass which converts dihydroxyacetonephosphate into pyruvate.

This bypass (Figure 5.11) is present in *Escherichia coli*, *Pseudomonas saccharophila*, and possibly in a number of other aerobic bacteria. Its physiological significance lies in the fact that it makes the formation of acetyl-CoA from dihydroxyacetonephosphate possible under conditions where low phosphate concentrations limit the activity of glyceraldehyde-3-phosphate dehydrogenase. In accordance with this role methylglyoxal synthase is inhibited by inorganic phosphate.

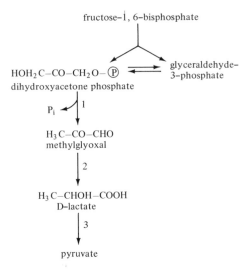

Figure 5.11. The methylglyoxal bypass. 1, methylglyoxal synthase; 2, glyoxalase I and II; 3, D-lactate oxidase (flavin-linked).

V.   Diversity in Energy Metabolism

After having seen that there are alternate routes of sugar degradation present in certain bacteria besides the Embden–Meyerhof pathway, one might then ask whether all aerobic heterotrophs use the same type of enzyme systems to synthesize ATP. As will be seen, there are differences here also.

A.   Pyruvate dehydrogenase

Pyruvate dehydrogenase is usually not present in anaerobes. It is a characteristic enzyme of aerobic heterotrophs, occurring in practically all aerobic bacteria that catabolize substrates via pyruvate. However, certain aerobes are very much restricted with respect to the nature of substrates they are able to utilize. *Hyphomicrobium* species, for instance, utilize C_1 and C_2 compounds only (methanol, methylamine, acetate); they lack pyruvate dehydrogenase. The same is probably true for other microorganisms limited to growth on C_1 compounds.

B.   Tricarboxylic acid cycle (TCA cycle)

There is no doubt that the great majority of aerobic heterotrophs use the TCA cycle in order to provide $NADH_2$ for the respiratory chain and precursors for the biosynthesis of cellular constituents. Some aerobes lack a complete cycle: a number of acetic acid bacteria and bacteria that are limited to growth on C_1 compounds (see Chapter 6). These organisms are devoid of α-oxoglutarate dehydrogenase. The cycle is thus interrupted but glutamate can still be synthesized via citrate, isocitrate, and α-oxoglutarate. Succinyl-CoA and compounds derived thereof are formed by these organisms from oxaloacetate via L-malate, fumarate, and succinate. Whereas the reduction of oxaloacetate to L-malate and the dehydration of the latter to fumarate are catalyzed by the corresponding TCA cycle enzymes, malate dehydrogenase and fumarase, a special enzyme is employed for succinate formation: fumarate reductase. It uses $NADH_2$ as hydrogen donor and thus is different from succinate dehydrogenase. Finally, succinyl-CoA is formed by succinate thiokinase.

C. Respiratory chain

Although the possession of a respiratory chain is a characteristic feature of all aerobes, differences have been encountered when the chains of a number of bacteria were analyzed for their components. This is in contrast to the respiratory chain of mitochondria, which does not show species-specific variations in its general composition. A few examples are given in Figure 5.12.

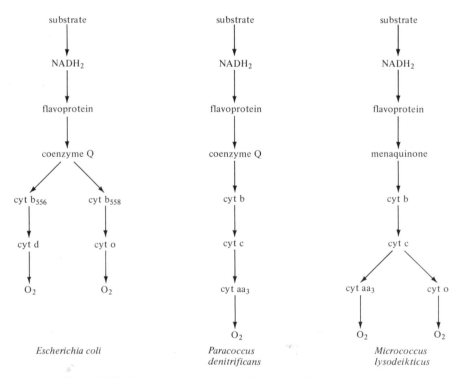

Figure 5.12. Components of the respiratory chain of three bacteria.

It can be seen that the chain of *E. coli*, which lacks a cytochrome of the c type, is not characteristic of all bacteria. The chains of *Paracoccus denitrificans* and *Micrococcus lysodeikticus* contain cytochromes of the b, c, and a type. It is not yet certain that the P/O ratio is the same in all bacteria. Experimental data indicate that the chain of *E. coli* contains two phosphorylation sites, but those of *P. denitrificans* and *M. lysodeikticus* have three

A remarkable property of the electron transport system of many bacteria is that under certain conditions nitrate can be used as H-acceptor instead of oxygen. This will be discussed in the next section.

VI. Dissimilatory Reduction of Nitrate

$$\text{glucose} + 6O_2 \longrightarrow 6CO_2 + 6H_2O \qquad \Delta G'_0 = -686 \text{ kcal} \quad (5.1)$$
$$(-2870\text{kJ})$$

$$\text{glucose} + 4.8NO_3^- + 4.8H^+ \longrightarrow$$
$$6CO_2 + 2.4N_2 + 8.4H_2O \qquad \Delta G'_0 = -638 \text{ kcal} \quad (5.2)$$
$$(-2669\text{kJ})$$

$$\text{glucose} + 12NO_3^- \longrightarrow 6CO_2 + 6H_2O + 12NO_2^- \qquad \Delta G'_0 = -422 \text{ kcal}$$
$$(-1766\text{kJ})$$
$$(5.3)$$

From the above equations it is apparent that the free energy changes of glucose oxidation with either nitrate or oxygen are comparable. Thus it is reasonable that certain bacteria are able to respire with nitrate instead of oxygen. In the course of this respiration they reduce nitrate to N_2 (and N_2O in some cases). These so-called **denitrifying bacteria** encompass a fairly large group of organisms of a great variety of genera, some of which are listed in Table 5.3. Especially, many bacilli and pseudomonads are able to carry out a denitrification.

Table 5.3. Denitrifying bacteria

Alcaligenes faecalis
Bacillus licheniformis
Hyphomicrobium vulgare
Paracoccus denitrificans
Pseudomonas stutzeri
Spirillum itersonii
Thiobacillus denitrificans

A number of bacteria, for instance most enterobacteria, are able to perform a **nitrate-nitrite respiration** in which nitrate is reduced to nitrite (Equation 5.3). The latter is either excreted or reduced by non-ATP-yielding reactions to ammonia (Figure 5.13).

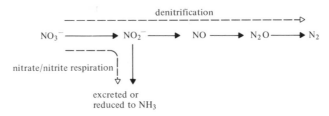

Figure 5.13. Denitrification and nitrate/nitrite respiration.

The enzyme machinery for both processes, nitrate/nitrite respiration and denitrification, is formed only under anaerobic conditions or conditions of low oxygen tension. In most cases nitrate is required as inducer. Also, the activity of the enzymes involved in dissimilatory nitrate reduction is strongly inhibited by oxygen. Thus, denitrification and nitrate/nitrite respiration take place only when oxygen is absent or available in insufficient amounts. All bacteria capable of performing these kinds of respiration prefer oxygen respiration if possible.

Like oxygen respiration, denitrification allows a complete oxidation of the organic substrate to CO_2. For instance, when *Bacillus licheniformis* grows with glucose and nitrate under anaerobic conditions, the substrate is degraded via the Embden–Meyerhof pathway; the acetyl-CoA formed from pyruvate is oxidized via the tricarboxylic acid cycle, and $NADH_2$ and $FADH_2$ thus formed serve as electron donors for the respiratory chain. Nitrate, however, does not simply replace oxygen; special types of cytochromes and membrane-bound enzyme systems are formed, which reduce nitrate to nitrite and further to nitrogen. The electron flow in denitrification is illustrated schematically in Figure 5.14.

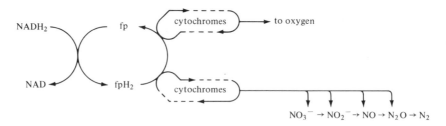

Figure 5.14. Electron flow in denitrification. fp, flavoprotein.

As is shown in Figure 5.14 four reduction steps can be distinguished in denitrification.

Nitrate reductase is a molybdenum-containing membrane-bound enzyme which reduces nitrate to nitrite.

$$NO_3^- + 2e^- + 2H^+ \longrightarrow NO_2^- + H_2O$$

Nitrite and **nitric oxide reductases** appear as soluble enzymes when cells of denitrifying bacteria are fractionated.

$$NO_2^- + e^- + H^+ \longrightarrow NO + OH^-$$
$$2NO + 2e^- + 2H^+ \longrightarrow N_2O + H_2O$$

Nitrous oxide reductase is membrane-bound.

$$N_2O + 2e^- + 2H^+ \longrightarrow N_2 + H_2O$$

Denitrifying bacteria can use nitrate—and many of them also nitrite, NO, and N_2O—as electron acceptors. This means that at least two of the four reduction steps are coupled with electron transport phosphorylation in the respiratory chain (NO_3^- and N_2O reduction). There is evidence that the reduction of NO_2^- to N_2O is also coupled to ATP formation.

In the nitrate/nitrite respiring bacteria only the reduction of nitrate to nitrite proceeds via the respiratory chain. Nitrite is excreted or reduced to ammonia by enzyme systems not involving the respiratory chain. Therefore, these bacteria gain ATP only from the reduction of nitrate to nitrite.

In this connection it should be mentioned that most nondenitrifying bacteria are also able to reduce nitrite and nitrate to ammonia by soluble enzyme systems, which are not oxygen-sensitive and are not coupled to ATP-yielding reactions. This process is called **assimilatory nitrate reduction**, and it has the purpose of supplying the cell with ammonia for the biosynthesis of cellular material. It is present in all bacteria that grow with nitrate as nitrogen source. In accordance with this, the formation of the assimilatory enzymes is repressed by ammonia.

VII. Alternate Anaplerotic Sequences

We have seen that the kind of anaplerotic sequence employed by *E. coli* depends on the growth substrate: on glucose, oxaloacetate is synthesized from phospho-enolpyruvate with PEP carboxylase; on acetate, the glyoxylate cycle is induced and PEP is formed with PEP carboxykinase; on pyruvate, the enzyme PEP synthetase is responsible for the provision of PEP for gluconeogenesis. Although not many bacteria have been studied in such detail as *E. coli*, the experimental results show that alternate anaplerotic sequences occur among bacteria also.

For substrates that are catabolized via PEP (glucose, other carbohydrates) the following two anaplerotic sequences are widespread among bacteria.

A. PEP carboxylase

$$PEP + HCO_3^- \longrightarrow \text{oxaloacetate} + P_i$$

This enzyme is responsible for the replenishment of oxaloacetate in enterobacteria, *Bacillus anthracis*, *Acetobacter xylinum*, *Thiobacillus novellus*, and *Azotobacter vinelandii*.

B. Pyruvate carboxylase

$$\text{pyruvate} + HCO_3^- + ATP \rightleftarrows \text{oxaloacetate} + ADP + P_i$$

Like mammalian systems a number of bacteria employ pyruvate carboxylase for oxaloacetate synthesis. This enzyme contains biotin, and the latter is directly involved in this carboxylation reaction:

$$\text{enzyme-biotin} + ATP + HCO_3^- \underset{Mg^{2+}}{\overset{\text{acetyl-CoA}}{\rightleftarrows}} \text{enzyme-biotin} \sim CO_2 + ADP + P_i$$

$$\text{enzyme-biotin} \sim CO_2 + \text{pyruvate} \rightleftarrows \text{enzyme-biotin} + \text{oxaloacetate}$$

enzyme - biotin ~ CO$_2$

Pyruvate carboxylase from mammalian sources requires acetyl-CoA for maximal activity. With the exception of *Pseudomonas citronellolis* and *P. aeruginosa*, this is also true for the enzyme from a number of bacteria including *Arthrobacter globiformis*, *Bacillus coagulans*, *Acinetobacter calcoaceticus*, and *Rhodopseudomonas sphaeroides*.

Bacteria which grow on acetate or on compounds channeled into the intermediary metabolism via acetyl-CoA (butyrate, aliphatic hydrocarbons) usually employ the glyoxylate cycle to form C_4-dicarboxylic acids and PEP carboxykinase to synthesize PEP for gluconeogenesis. Certain phototrophic bacteria (*Rhodospirillum rubrum*, *Rhodopseudomonas sphaeroides*), however, contain malate synthase during aerobic growth on acetate but lack isocitrate lyase. It is not yet known how these microorganisms form glyoxylate.

Many bacteria are able to grow on pyruvate and lactate. Those that contain pyruvate carboxylase do not have problems with respect to oxaloacetate synthesis. The others—as we have already seen for *Escherichia coli*— may use **phospho-enolpyruvate synthetase** to form PEP from pyruvate and **phospho-enolpyruvate carboxylase** to replenish the oxaloacetate pool. Not

Table 5.4. Kind of enzyme systems involved in oxaloacetate and PEP synthesis, as dependent on the nature of the growth substrate

catabolic sequence	synthesis of	
	oxaloacetate	PEP
substrate ↓ PEP ↓ pyruvate ↓ acetyl-CoA	PEP carboxylase or pyruvate carboxylase	
substrate ↓ pyruvate ↓ acetyl-CoA	pyruvate carboxylase or PEP carboxylase	PEP carboxykinase or PEP synthetase or pyruvate-phosphate dikinase
substrate ↓ acetyl-CoA	glyoxylate cycle	PEP carboxykinase

much is known about the distribution of the synthetase. It is present in *E. coli* and *Salmonella typhimurium* and presumably in other enterobacteria. *Acetobacter xylinum* and the anaerobe *Propionibacterium shermanii* contain another enzyme that also allows the direct synthesis of PEP from pyruvate: **pyruvate-phosphate dikinase**. The reactions catalyzed by PEP synthetase and the dikinase are given in the following equations:

$$\text{pyruvate} + \text{ATP} \underset{\xleftarrow{\hspace{2cm}}}{\overset{\text{PEP synthetase}}{\xrightarrow{\hspace{2cm}}}} \text{PEP} + \text{AMP} + \text{P}_i$$

$$\text{pyruvate} + \text{ATP} + \text{P}_i \underset{\xleftarrow{\hspace{2cm}}}{\overset{\text{pyruvate-phosphate dikinase}}{\xrightarrow{\hspace{2cm}}}} \text{PEP} + \text{AMP} + \text{PP}_i$$

The differences are apparent. If pyrophosphorylated enzymes are intermediates, the PEP synthetase releases one of these phosphate groups as phosphate and transfers the second to pyruvate; in the dikinase reaction one of the phosphate groups is transferred to phosphate to yield pyrophosphate and the second to pyruvate.

Table 5.4 summarizes the function of the enzyme systems discussed in this section.

VIII.   Biosynthesis of Monomers and Polymers

The pathways used by microorganisms to synthesize amino acids, purine, and pyrimidine bases, the various lipids, and carbohydrates are usually the same as those outlined for *Escherichia coli*. There is not much diversity as to the nature of the intermediates involved in the synthesis of these compounds. However, this does not mean that the enzymes catalyzing the same type of reaction in different bacteria are identical. For example, chorismate mutase and aspartokinase of *E. coli* are different from the corresponding enzymes of *Bacillus subtilis*. Their molecular weight and chemical and physical properties are different, but in both cases the enzymes are indeed either chorismate mutase or aspartokinase. The same is true for many enzymes of different bacteria. It is noteworthy that enzymes specific for the same reaction very often have regulatory properties that differ from one bacterial species to the other. This will be discussed in Chapter 7.

We can therefore conclude that there is a relatively constant set of anabolic reactions in bacteria, on the one hand, and much variation in the structure and the properties of the enzyme systems involved, on the other.

Many bacteria are, however, devoid of enzyme systems required for the synthesis of one or more precursors of polymers. *E. coli*, some other enterobacteria, *Bacillus megaterium*, and *Azotobacter vinelandii*—to mention just a few—are able to grow with glucose and minerals, that is, they contain the entire enzymatic machinery necessary to make all compounds required for growth. A number of bacilli and pseudomonads depend for growth on the presence of certain vitamins and amino acids in their environment. In the laboratory they are grown in media supplemented with peptone and yeast or beef extract as sources of these compounds. Particularly dependent on preformed monomers are bacteria that normally grow on complex organic material (juice, milk, decaying plants, etc.). The anaerobic lactic acid bacteria, for instance, are so poorly equipped with anabolic enzyme systems that they require practically all the monomers for growth.

As far as is known, the synthesis of informational macromolecules (DNA, RNA, and proteins) proceeds in all bacteria in the same fashion. This is also true for periodic macromolecules, although one has to consider here that not all bacteria form periodic macromolecules of the same composition and structure.

Bacteria differ in the type of **reserve material** they accumulate under certain conditions. Table 5.5 lists a small selection of bacteria that accumulate one or more of the typical storage materials of microorganisms: glycogen, poly-β-hydroxybutyrate, or polyphosphate. There seem to be some microorganisms that do not form any reserve material; one of these is *Pseudomonas aeruginosa*.

Polyphosphate accumulation is widespread among bacteria. It functions as a phosphorus storage material and is utilized for nucleic acid and phospholipid synthesis under conditions of phosphorus starvation.

Table 5.5 Occurrence of glycogen, poly-β-hydroxybutyrate (PHB), and polyphosphate in bacteria

organism	glycogen	PHB	polyphosphate
Enterobacter aerogenes	+	−	+
Alcaligenes eutrophus	−	+	+
Azotobacter vinelandii	−	+	+
Bacillus megaterium	+	+	−
Escherichia coli	+	−	+
Mycobacterium phlei	+	−	+
Pseudomonas aeruginosa	−	−	−
Pseudomonas multivorans	−	+	?
Rhodospirillum rubrum	+	+	?
Sphaerotilus natans	−	+	?

Glycogen and **poly-β-hydroxybutyrate** serve as energy-storage compounds. In the absence of an exogenous energy source they are slowly degraded and supply the cells with ATP for maintenance metabolism. The enzymes involved in glycogen synthesis and degradation were described in Chapter 3.

Poly-β-hydroxybutyrate (PHB) is a typical prokaryotic storage material. It is widespread in bacilli, in chemolithotrophic and phototrophic bacteria, and in pseudomonads (but does not occur in the fluorescent group). PHB is a polymer of $D(-)$-β-hydroxybutyrate and has a molecular weight between 60,000 and 250,000.

$$\text{HO}-\underset{\underset{\text{CH}_3}{|}}{\text{CH}}-\text{H}_2\text{C}-\underset{\underset{\text{O}}{\|}}{\text{C}}-\left[-\text{O}-\underset{\underset{\text{CH}_3}{|}}{\text{CH}}-\text{CH}_2-\underset{\underset{\text{O}}{\|}}{\text{C}}-\right]_n-\text{O}-\underset{\underset{\text{CH}_3}{|}}{\text{CH}}-\text{CH}_2-\text{COOH}$$

poly-β-hydroxybutyric acid

It is accumulated in the cells as granules surrounded by membranes. Under appropriate conditions (no nitrogen source but plenty of energy and organic substrates) bacterial cells may accumulate so much PHB that it accounts for approximately 60% of their dry weight.

Synthesis of PHB in *Azotobacter beijerinckii* proceeds as shown in Figure 5.15. The polymerase appears to be granule-bound. An alternate route is employed by *Rhodospirillum rubrum* to synthesize the $D(-)$-monomer. This microorganism contains a $L(+)$-β-hydroxybutyryl-CoA dehydrogenase but in addition two crotonases that bring about the conversion of the $L(+)$ form into the $D(-)$ form:

$L(+)$-β-hydroxybutyryl-CoA

H_2O crotonase I

crotonyl-CoA

H_2O crotonase II

$D(-)$-β-hydroxybutyryl-CoA

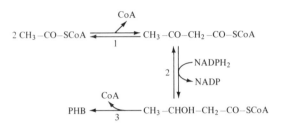

Figure 5.15. Synthesis of PHB in *Azotobacter beijerinckii*. 1, β-ketothiolase; 2, D(−)-β-hydroxybutyryl-CoA dehydrogenase; 3, D(−)-β-hydroxybutyryl-CoA polymerase.

Utilization of PHB starts when an exogenous energy source is no longer available to the cells. A depolymerase then releases D(−)-β-hydroxybutyrate from the granules, which is subsequently oxidized to acetoacetate by a NAD-specific D(−)-β-hydroxybutyrate dehydrogenase. Acetoacetate is then channeled into the intermediary metabolism by a CoA transferase reaction (Figure 5.16). Thus, CoA esters are the substrates for PHB synthesis and acids are the products of PHB degradation; this separation facilitates the regulation of both processes.

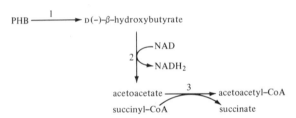

Figure 5.16. Degradation of PHB. 1, depolymerase; 2, D(−)-β-hydroxybutyrate dehydrogenase; 3, CoA transferase.

In a discussion of the diversity of pathways involved in periodic macro-molecule biosynthesis, it is also necessary to consider that bacteria vary with respect to the composition of their **cell walls**. This implies, of course, that different enzyme systems are involved in the formation of these structures in different bacteria.

The synthesis of the layers of *E. coli* cell wall has been outlined in Chapter 3. Approximately 10% of the wall of *E. coli* and of other Gram-negative bacteria consists of peptidoglycan. The investigation of the walls of a fairly large number of Gram-negative bacteria has shown that the composition of their peptidoglycan (murein) layers is very similar (Gram-negative bacteria investigated, for instance, include *Acetobacter xylinum*, *Aerobacter aerogenes*, *Azotobacter chroococcum*, *Caulobacter*, *E. coli*, *Proteus mirabilis*, *Pseudomonas aeruginosa*, and *Spirillum serpens*). They all contain N-acetylmuramic acid, N-acetylglucosamine, 2,6-diaminopimelate, alanine, and glutamate in molar ratios of 1:1:1:2:1. Cross-linkage of the peptide chains always occurs between the amino group of diaminopimelate and D-alanine (see Figure 3.25).

The outer membrane of Gram-negative bacteria is quite variable in its composition. A number of different phospholipids are found there and the nature of the sugar moieties in the lipopolysaccharide fraction varies even within one species (various serotypes of *Salmonella* according to the Kaufmann–White scheme). Clearly, Gram-negative bacteria must contain species-specific enzyme systems, which bring about these differences in the composition of their cell walls.

Gram-positive bacteria contain a very complex peptidoglycan as the major component of the cell wall. In many of these microorganisms it accounts for 80 to 90% of the cell wall components. As in the peptidoglycan of Gram-negative bacteria, its backbone consists of alternating sequences of N-acetylmuramic acid and N-acetylglucosamine. However, according to the work of Kandler, Schleifer, and others the peptidoglycan of Gram-positive bacteria differs from that of Gram-negative bacteria in two respects:

1. The composition of the tetrapeptide connected with the backbone is variable to a certain extent. Table 5.6 gives a few examples. We find the greatest variability in position 3. Whereas the Gram-negative bacteria always contain meso-diaminopimelate in this position, in many Gram-positive bacteria it is replaced by either L-lysine, L-ornithine, or homoserine. Also in some of the tetrapeptides the α-carboxyl group of D-glutamate is amidated (the γ-carboxyl group always forms the peptide bond with the subsequent amino acid). Finally, the peptidoglycan of *Eubacterium limosum* contains L-serine instead of L-alanine as the first amino acid of the tetrapeptide.
2. In contrast to the Gram-negative bacteria, in which peptide chains are always connected by a peptide bond between the ε-amino group of 2,6-diaminopimelate and the carboxyl group of D-alanine, cross-linkage in

Table 5.6. Composition of tetrapeptides of peptidoglycans of Gram-positive bacteria[a]

tetrapeptide	organisms
L-Ala-D-Gln-L-Lys-D-Ala	lactobacilli, streptococci
L-Ala-D-Glu-meso-DAP-D-Ala	bacilli
L-Ala-D-Gln-L-Orn-D-Ala	some micrococci
L-Ala-D-Glu-Homoser-D-Ala	some corynebacteria
L-Ser-D-Glu-L-Orn-D-Ala	*Eubacterium limosum*

[a]D. A. Reaveley and R. E. Burge, *Adv. Microbial Physiol.* **7**, 2–71 (1972).

Gram-positive bacteria is accomplished in different ways:

(a) Direct cross-linkage between positions 3 and 4 of two tetrapeptides (most bacilli, like in Gram-negative bacteria).

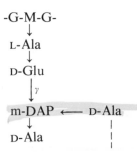

(b) Cross-linkage between positions 3 and 4 by interpeptide bridges such as penta-glycyl chains (staphylococci) or chains (2 to 5 amino acids) containing L-Ala, L-Thr, L-Ser, and/or Gly (*Arthrobacter*, micrococci, streptococci). *Cellulomonas* species contain an aspartate or glutamate bridge.

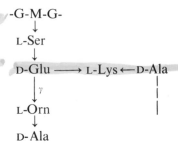

(c) Cross-linkage between the α-carboxyl group of D-glutamate at position 2 of one peptide and the carboxyl group of D-alanine of the second peptide, via a diamino acid (Lys, Orn). This type of linkage is found in *Eubacterium limosum* and in some corynebacteria.

-G-M-G-
↓
L-Ser
↓
D-Glu ⟶ L-Lys ←D-Ala
│γ   │
L-Orn   │
↓
D-Ala

It is apparent that there is a great variability within the group of Gram-positive bacteria as to the composition of their peptidoglycans.

Gram-negative bacteria contain a very thin peptidoglycan layer (mono- or bimolecular layer). In Gram-positive bacteria, this layer is very thick and cross-linkage occurs not only between adjacent chains but also between layers above and below one another.

The cell wall of Gram-positive bacteria does not contain components comparable to the outer membrane fraction of the Gram-negative bacteria. However, in addition to the peptidoglycan layer the wall contains various proteins and the so-called **teichoic acids**, which account for about 10 to 20% of its dry weight. Teichoic acids are polymers of glycerol or ribitolphosphate substituted with various sugars and with D-alanine.

```
HO—CH₂                    O—CH₂                    O—CH₂
   |                         |                         |
H—C—O Ala      O        H—C—O Ala      O        H—C—O Ala
   |          ‖            |            ‖            |
H—C—OH        P         H—C—OH         P         H—C—OH
   |          |            |            |            |
H—C—O sugar   OH        H—C—O sugar     OH        H—C—O sugar
   |                         |                         |
  CH₂O                      CH₂O                      CH₂O
```

1,5-poly(ribitolphosphate) teichoic acid

They are covalently bound to the peptidoglycan layer by phosphate ester bridges. One function of the teichoic acids is to maintain a high concentration of divalent cations in the vicinity of the cells. Moreover, they contribute to the stability of the walls of Gram-positive bacteria.

IX. Summary

1. When comparing the mode in which substances from the environment cross the cell membrane into the cytoplasm, four different mechanisms can be distinguished: passive diffusion, facilitated diffusion, active transport, and group translocation. The latter two processes are important for the uptake of substrates by bacteria.

Active transport allows transport against a concentration gradient. The energy required is normally provided by the energized membrane, in some systems directly by ATP. The process of group translocation is very efficient because the substrate is trapped by chemical modification in the membrane. Glucose is converted to the 6-phosphate and fructose to the 1-phosphate. The corresponding phosphotransferase systems are used predominantly by facultative and strict anaerobic bacteria. Many aerobes use active transport systems for sugars.

2. Pseudomonads and many other Gram-negative bacteria employ the Entner–Doudoroff pathway for the breakdown of hexoses. The key enzymes of this pathway are 6-phosphogluconate dehydratase and 2-keto-3-deoxy-6-phosphogluconate (KDPG) aldolase. The conversion of glucose into 2 pyruvate via the ED pathway yields 1 ATP whereas 2 ATP are formed in the EM pathway. The ED pathway is also important for bacteria growing with gluconate, mannonate, and hexuronates.

3. In the oxidative pentose phosphate cycle, glucose is oxidized to 3 CO_2 and glyceraldehyde-3-phosphate. Subsequent oxidation of the latter in the tricarboxylic acid cycle results in the breakdown of glucose without participation of the ED and EM pathways.

4. At low concentrations of phosphate, a number of bacteria employ the methylglyoxal bypass to form acetyl-CoA from dihydroxyacetonephosphate.

5. The tricarboxylic acid cycle is employed by most aerobic heterotrophs

for the generation of reducing power. Organisms limited to the metabolism of C_1 compounds and some acetic acid bacteria do not contain a complete cycle; they are devoid of α-oxoglutarate dehydrogenase. These bacteria can synthesize L-glutamate via citrate and succinyl-CoA via fumarate and succinate.

The compositions of the respiratory chains of bacteria are different. Variations are found in number and type of the cytochromes involved.

6. A number of bacteria are able to respire with nitrate instead of oxygen as H-acceptor. Nitrate respiration proceeds only under anaerobic conditions. Two processes are distinguished: denitrification in which nitrate is reduced to N_2 or N_2O and nitrate/nitrite respiration in which nitrate is reduced to nitrite. The latter is excreted or reduced to ammonia.

7. Bacteria which metabolize substrates via PEP use either PEP carboxylase or pyruvate carboxylase in order to synthesize oxaloacetate. The glyoxylate cycle is generally employed for oxaloacetate synthesis when aerobes grow on acetate. PEP synthetase and pyruvate-phosphate dikinase serve several microorganisms for PEP synthesis from pyruvate.

8. Reserve materials accumulated by bacteria under certain conditions are polyphosphate and the energy-storage compounds glycogen and poly-β-hydroxybutyrate.

9. Bacteria vary with respect to the composition of their cell walls. In Gram-negative bacteria, the composition of the peptidoglycan layer is very similar but the outer membrane layer shows much variation as to the nature of phospholipids and sugars present.

In Gram-positive bacteria, the peptidoglycan layer accounts for 80 to 90% of the cell wall components. The composition of the tetrapeptide chains and the type of cross-linkages vary among the Gram-positive bacteria. L-Lysine and L-ornithine are found instead of m-diaminopimelate at position 3 of the peptide chains; besides the direct cross-linkage between positions 3 and 4 of two adjacent peptides, interpeptide bridges such as pentaglycyl bridges are found. Cell walls of Gram-positive bacteria also contain proteins and teichoic acids.

Chapter 6
Catabolic Activities of Aerobic Heterotrophs

Thus far we have discussed aerobic growth on very common substrates, such as glucose, lactose, or some organic acids. Clearly, mineralization of organic matter requires that microorganisms have the ability to degrade a vast number of organic compounds. When animals and plants die, a number of low-molecular-weight compounds become available to the microorganisms, as well as polymers, such as starch, cellulose, other polysaccharides, nucleic acids, and proteins. Furthermore, end products of bacterial fermentations (methane, propionate, butyrate, etc.) diffuse into aerobic zones and can serve as growth substrates.

Many bacteria exhibit an enormous flexibility as to the nature and the number of substrates they can utilize. Den Dooren de Jong found that 200 different organic compounds can serve as sole source of carbon and energy for *Pseudomonas putida*. Other pseudomonads, *Bacillus*, *Azotobacter*, and *Acinetobacter* species are also remarkable in that they can grow with many substrates. On the other hand, there are organisms that are more "specialized"; *Bacillus fastidiosus*, for instance, can utilize only uric acid and related purine compounds. Many bacteria growing with methane are restricted to C_1-compounds as growth substrates. Some of the catabolic activities of aerobic microorganisms will be discussed in this chapter.

I.   Degradation of Polymers by Exoenzymes

Polymers as such cannot penetrate the cell membrane. Therefore, bacteria excrete enzymes (in most cases hydrolases) that degrade the polymers to small transportable molecules.

Starch and **glycogen** [$\alpha(1 \rightarrow 4)$ glycosidic linkages; $\alpha(1 \rightarrow 6)$ linkages at branch points] are degraded by amylases. **α-Amylases** produced by bacilli and pseudomonads cleave starch first into dextrins and subsequently into maltose, glucose, and oligosaccharides with the $(1 \rightarrow 6)$ linkages left intact (Figure 6.1).

β-Amylase acts in a different way; it successively removes maltose residues from the nonreducing end of starch or glycogen. This type of enzyme is common in plants but not in bacteria.

The $\beta(1 \rightarrow 4)$ glycosidic linkages of **cellulose** are hydrolyzed by **cellulase**. This enzyme is excreted by a number of bacteria (*Cytophaga*, *Cellulomonas*, rumen bacteria such as *Bacteroides* and *Ruminococcus* species, some bacilli, and clostridia) and by many fungi (e.g., *Trichoderma viride*, *Aspergillus niger*). Some of the cellulases produced are not true exoenzymes. The cellulytic activity of the *Cytophaga* species, for instance, remains associated with the cell surface. Products of the enzymatic hydrolysis of cellulose are glucose and **cellobiose**, which is a disaccharide with a $\beta(1 \rightarrow 4)$ glycosidic linkage. Cellobiose is a potent inhibitor of cellulase activity. However, it is rapidly taken up by the cells and degraded.

The intracellular breakdown of disaccharides by bacteria is often initiated by a **phosphorylytic cleavage**:

$$\text{cellobiose} + \text{P}_i \xrightarrow{\text{cellobiose phosphorylase}} \text{glucose-1-P} + \text{glucose}$$

$$\text{maltose} + \text{P}_i \xrightarrow{\text{maltose phosphorylase}} \text{glucose-1-P} + \text{glucose}$$

$$\text{sucrose} + \text{P}_i \xrightarrow{\text{sucrose phosphorylase}} \text{glucose-1-P} + \text{fructose}$$

The phosphorylases are, of course, more economical than hydrolases such as β-galactosidase and invertase (hydrolyzes sucrose to fructose + glucose). The energy of the glycosidic link is saved and ATP is not required for the formation of sugar-1-phosphate.

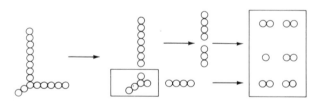

Figure 6.1. Action of α-amylase. Formation of maltose (∞), glucose ($\circ$), and the oligosaccharide with the $(1 \rightarrow 6)$ linkage.

Sucrose phosphorylase was first discovered in *Pseudomonas saccharophila*. Maltose and cellobiose phosphorylases occur in bacteria that decompose starch and cellulose.

Other polysaccharides and related compounds are also hydrolyzed by specific exoenzymes. Many bacteria are able to produce **pectinase** and to degrade pectin (e.g., *Bacillus polymyxa*, *Erwinia carotovora*, *Clostridium felsineum*). Corynebacteria, *Chromobacterium violaceum*, *Pseudomonas chitinovorans*, and other soil bacteria excrete **chitinase**. Other exoenzymes of bacterial origin are **hyaluronidase**, **neuraminidase**, **xylanase**, and **agarase**.

In order to be able to utilize the proteins of dead organisms many bacteria excrete **proteases**; these include bacilli, pseudomonads, *Proteus vulgaris*, and also many anaerobes. A well-characterized protease is **subtilisin** produced by *Bacillus subtilis*. It is not specific for peptide bonds between certain amino acids and it will hydrolyze both internal and terminal peptide bonds.

Nucleic acids are hydrolyzed by **ribonucleases** (RNases) and **deoxyribonucleases** (DNases). The excretion of the latter type of enzyme has been reported for hemolytic streptococci, *Staphylococcus aureus*, and clostridial species. A powerful ribonuclease is excreted by *Bacillus subtilis* in the late log phase of growth and in the stationary phase. The extracellular RNase has a lower molecular weight than the intracellular enzyme of *B. subtilis*. Both also differ in specificity; while the intracellular RNase is unspecific, the exoenzyme preferentially cleaves phosphodiester bonds of 3′-guanine

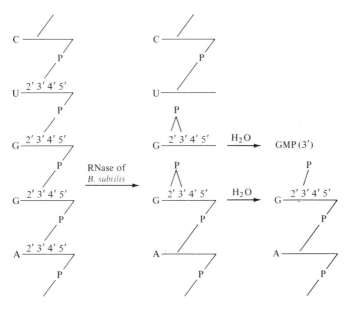

Figure 6.2. Action of extracellular RNase specific for phosphodiester bonds of 3′-guanine nucleotides. [F. Egami and K. Nakamura, *Microbial Ribonucleases*. Springer-Verlag, Berlin-Heidelberg-New York, p. 29 (1969).]

nucleotides such that 2′,3′-cyclic nucleotides are intermediates and 3′-phosphates the products (Figure 6.2). Other extracellular RNases are purine-specific (hydrolyze bonds of 3′-adenine and guanine nucleotides).

II.   Growth with Amino Acids

Amino acids and low-molecular-weight peptides produced by proteases are actively taken up and utilized for growth by many microorganisms. Some of the amino acids are structurally so related to central intermediates of cell metabolism that their degradation is very easy. In most cases the amino acid is first converted to the corresponding keto acid.

$$\begin{aligned}
\text{glutamate} &\longrightarrow \text{2-oxoglutarate}\\
\text{aspartate} &\longrightarrow \text{oxaloacetate}\\
\text{alanine} &\longrightarrow \text{pyruvate}\\
\text{valine} &\longrightarrow \text{2-oxoisovalerate}\\
\text{leucine} &\longrightarrow \text{2-oxoisocaproate}\\
\text{isoleucine} &\longrightarrow \text{2-oxo-3-methylvalerate}
\end{aligned}$$

This **oxidative deamination** can be accomplished in the following different ways.

(a)   Oxidation by cytochrome-linked oxidases

$$R-\underset{\underset{NH_2}{|}}{CH}-COOH + \tfrac{1}{2}O_2 \longrightarrow R-CO-COOH + NH_3$$

Several bacteria contain L-amino and D-amino acid oxidases. The latter are important because of the presence of D-amino acids in some polymers (e.g., peptidoglycan) and because they work together with racemaces, which catalyze the conversion of L-amino acids into D-amino acids. The oxidases are flavoproteins and feed the electrons into the respiratory chain. They are relatively unspecific, and a particular oxidase may attack 10 different amino acids.

(b)   Oxidation by NAD(P)-linked dehydrogenases

$$CH_3-\underset{\underset{NH_2}{|}}{CH}-COOH + NAD + H_2O \underset{\longleftarrow}{\overset{\text{alanine dehydrogenase}}{\longrightarrow}}$$
$$CH_3-CO-COOH + NADH_2 + NH_3$$

Alanine dehydrogenase occurs in a number of bacilli and clostridia. Glutamate dehydrogenase is very widespread and catalyzes the analogous reaction with glutamate as substrate.

(c)   Transamination with pyruvate or 2-oxoglutarate as acceptor of the amino group and subsequent regeneration of the acceptor by a dehydrogenation reaction (as in B).

$$
\underset{\substack{\text{HCNH}_2 \\ | \\ \text{COOH}}}{\overset{\text{R}}{|}} + \underset{\substack{\text{CO} \\ | \\ \text{COOH}}}{\overset{\text{CH}_3}{|}} \xrightleftharpoons[\text{alanine dehydrogenase}]{\text{transaminase}} \underset{\substack{\text{C}=\text{O} \\ | \\ \text{COOH}}}{\overset{\text{R}}{|}} + \underset{\substack{\text{HCNH}_2 \\ | \\ \text{COOH}}}{\overset{\text{CH}_3}{|}}
$$

Whereas 2-oxoglutarate, oxaloacetate, and pyruvate can be easily handled by bacterial cells, specific catabolic routes are required to channel 2-oxo-isovalerate or 2-oxoisocaproate into the intermediary metabolism. The routes found in bacteria are the same as those found in animals, and lead to the formation of acetyl-CoA and propionyl-CoA.

Another reaction used to initiate the breakdown of amino acids is **deamination**. This reaction with serine and threonine yields pyruvate and 2-oxobutyrate, respectively. 2-Oxobutyrate can be oxidized by a multienzyme complex resembling the pyruvate dehydrogenase complex to yield propionyl-CoA; the metabolic fate of the latter will be outlined later.

$$
\underset{\substack{| \quad\; | \\ \text{OH} \; \text{NH}_2 \\ \text{serine}}}{\text{CH}_2-\text{CH}-\text{COOH}} \xrightarrow{\substack{\text{serine} \\ \text{dehydratase}}} \underset{\text{pyruvate}}{\text{CH}_3-\text{CO}-\text{COOH}} + \text{NH}_3
$$

$$
\underset{\substack{| \quad\; | \\ \text{OH} \; \text{NH}_2}}{\text{H}_3\text{C}-\text{CH}-\text{CH}-\text{COOH}} \xrightarrow{\substack{\text{threonine} \\ \text{dehydratase}}} \underset{\text{2-oxobutyrate}}{\text{CH}_3-\text{CH}_2-\text{CO}-\text{COOH}} + \text{NH}_3
$$

These reactions start with the removal of water:

$$
\underset{\substack{| \\ \text{HC}-\text{NH}_2 \\ | \\ \text{COOH}}}{\overset{\text{CH}_2\text{OH}}{|}} \xrightarrow{\text{H}_2\text{O}} \underset{\substack{\| \\ \text{C}-\text{NH}_2 \\ | \\ \text{COOH}}}{\overset{\text{CH}_2}{|}} \xrightarrow[\text{H}_2\text{O}\qquad\text{NH}_3]{} \underset{\substack{| \\ \text{C}=\text{O} \\ | \\ \text{COOH}}}{\overset{\text{CH}_3}{|}}
$$

Catabolic threonine dehydratase of *E. coli* differs in its properties, especially in its regulatory properties, from the anabolic threonine dehydratase present in the same microorganism; the latter catalyzes the first step of a reaction sequence leading from 2-oxobutyrate to isoleucine.

Many aerobes synthesize aspartase when they grow on aspartate:

$$
\text{aspartate} \xrightleftharpoons{\text{aspartase}} \text{fumarate} + \text{NH}_3
$$

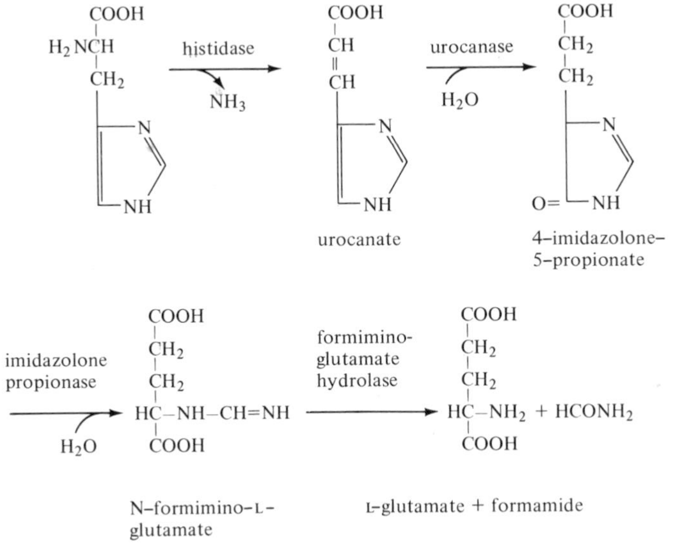

Figure 6.3. Breakdown of histidine.

A similar reaction initiates the breakdown of histidine (Figure 6.3). Aerobic breakdown of the aromatic amino acids is also feasible for many microorganisms. Thus it is understandable that many aerobes grow well on nutrient broth and other media containing mostly amino acids and peptides.

III.   Growth with Organic Acids

In connection with the general metabolism of *E. coli*, we have already discussed the metabolic implications of growth on acids such as pyruvate and acetate, particularly the requirement of special anaplerotic sequences for growth on these compounds. Some further remarks about growth on organic acids are necessary. A number of aerobes can grow on **long-chain fatty acids** (pseudomonads, *Acinetobacter*, bacilli, *E. coli*); they are usually degraded by **β-oxidation**. The fatty acid, for instance palmitic acid, is first converted to the corresponding coenzyme A ester by acyl-CoA synthetase. Such synthetases have a low specificity. The enzyme from *Bacillus megaterium*, for instance, reacts with acids from C_6 to C_{20}. The CoA ester is then oxidized in the β-position and subsequently cleaved to yield acetyl-CoA and the CoA ester of the fatty acid shortened by two carbon atoms. One β-oxidation cycle requires the action of four enzymes, as is illustrated in Figure 6.4. Even-numbered fatty acids yield only acetyl-CoA and it is, therefore, apparent that microorganisms growing on these substrates require the glyoxylate cycle as

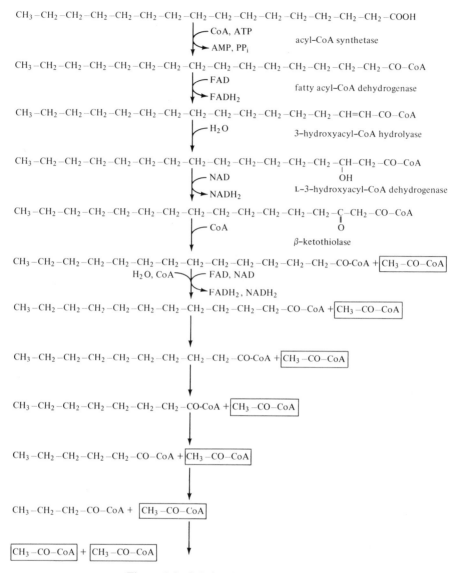

Figure 6.4. β-Oxidation of palmitic acid.

anaplerotic sequence and PEP carboxykinase to form PEP for gluconeogenesis.

With odd numbered fatty acids, the final β-oxidation cycle yields acetyl-
CoA and propionyl-CoA. The latter is also formed in valine and isoleucine
catabolism, and, in fact, a number of bacteria are able to grow with propionate as energy and carbon source.

The metabolism of **propionate** is initiated by its activation to propionyl-
CoA. For the further metabolism of propionyl-CoA several different path-

ways have been found which might be involved. The conversion of propionyl-CoA to succinyl-CoA is widespread; it occurs in animal tissues, *Paracoccus denitrificans*, and rhizobia. The first enzyme carboxylates propionyl-CoA to yield methylmalonyl-CoA, which undergoes rearrangement reactions leading to succinyl-CoA. The mutase catalyzing the last reaction contains a derivative of vitamin B_{12} as an essential coenzyme. The formation of succinyl-CoA is shown in Figure 6.5. Since these reactions are so important in the propionate fermentation, more details will be given in Chapter 8.

E. coli converts propionyl-CoA to pyruvate by the following reactions:

$$
\begin{array}{ccccccc}
CH_3 & & CH_2 & & CH_3 & & CH_3 \\
| & \xrightarrow{2H} & \| & \xrightarrow{H_2O} & | & \xleftarrow{CoA} & | \\
CH_2 & & CH & & CHOH & & C{=}O \\
| & & | & & | & \xrightarrow{2H} & | \\
CO{-}SCoA & & CO{-}SCoA & & CO{-}SCoA & & COOH \\
\text{propionyl-CoA} & & \text{acrylyl-CoA} & & \text{lactyl-CoA} & & \text{pyruvate}
\end{array}
$$

The same intermediates occur in *Moraxella lwoffi*, but they are enzyme-bound and are not found as CoA esters.

Glyoxylate appears in nature as a degradation product of the purine bases. Under aerobic conditions these bases are oxidized by xanthine oxidase to uric acid. The latter is oxidized further to allantoin by the enzyme uricase. Hydrolysis of allantoin yields allantoate, which in different microorganisms is metabolized further by one of the two reaction sequences illustrated in Figure 6.6. *Pseudomonas aeruginosa*, *P. fluorescens*, and *Penicillium* species form first 2 mol of urea and 1 mol of glyoxylate. *P. acidovorans*, *Streptococcus allantoicus*, *Arthrobacter allantoicus*, and *E. coli*, on the other hand, first remove ammonia and CO_2 from allantoate so that finally 1 mol of urea per mol of glyoxylate is formed. These differences are probably not important for the microorganisms because they synthesize the enzyme urease when ammonia becomes growth-limiting. Urease cleaves urea to ammonia and

$$
\begin{array}{ccc}
CH_3 & \overset{CO_2 \quad P_i}{\underset{biotin}{\overset{ATP \quad ADP}{\underset{1}{\rightleftharpoons}}}} & COOH \\
| & & | \\
CH_2 & & HC{-}CH_3 \qquad \xrightarrow{\quad 2 \quad} \\
| & & | \\
Co{-}SCoA & & CO{-}SCoA \\
\text{propionyl-CoA} & & \text{L-methylmalonyl-CoA}
\end{array}
$$

$$
\begin{array}{ccc}
COOH & & COOH \\
| & \xrightarrow[B_{12}]{\quad 3 \quad} & | \\
H_3C{-}C{-}H & & H_2C{-}CH_2 \\
{\cdot}{-}{-}CO{-}SCoA & & CO{-}SCoA \\
\text{D-methylmalonyl-CoA} & & \text{succinyl-CoA}
\end{array}
$$

Figure 6.5. Formation of succinyl-CoA from propionyl-CoA. 1, propionyl-CoA carboxylase; 2, methylmalonyl-CoA racemase; 3, methylmalonyl-CoA mutase.

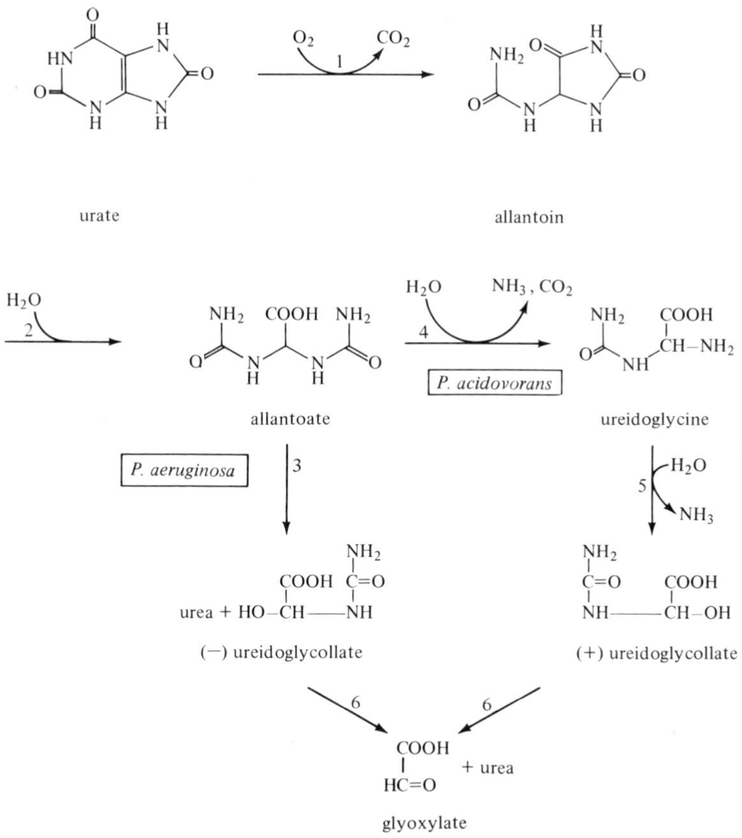

Figure 6.6. Degradation of urate to glyoxylate by *P. aeruginosa* and *P. acidovorans*. 1, uricase; 2, allantoinase; 3, allantoicase; 4, allantoate amidohydrolase; 5, ureidoglycine aminohydrolase; 6, ureidoglycolase.

carbon dioxide:

$$NH_2-C-NH_2 + H_2O \xrightarrow{\text{urease}} 2NH_3 + CO_2$$
$$\overset{\|}{O}$$

Growth of *E. coli* and pseudomonads on glyoxylate requires some enzymes that have not as yet been discussed. The provision of reducing power for the respiratory chain is very easy with this substrate. Glyoxylate is introduced into a cyclic process by the malate synthase reaction. The acceptor, acetyl-CoA, is regenerated by a series of reactions called the **dicarboxylic acid cycle** (Figure 6.7). In the course of oxidation of 1 mol of glyoxylate, 2 mol of NAD are reduced. For the replenishment of PEP and acetyl-CoA, which are intermediates of this cycle, an anaplerotic sequence is

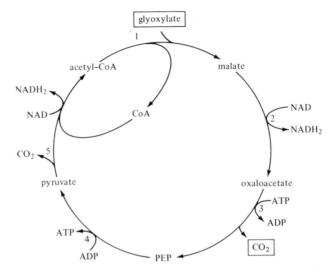

Figure 6.7. Oxidation of glyoxylate by the dicarboxylic acid cycle. 1, malate synthase;
2, malate dehydrogenase; 3, PEP carboxykinase; 4, pyruvate kinase; 5, pyruvate
dehydrogenase complex.

required. It consists of three enzymes and is known as the **glycerate pathway**
(Figure 6.8). The first enzyme, glyoxylate carboligase, catalyzes the condensation of two molecules of glyoxylate to CO_2 and tartronate semialdehyde.
The latter is reduced to glycerate, which is phosphorylated to yield 3-
phosphoglycerate—an intermediate of the Embden–Meyerhof pathway.

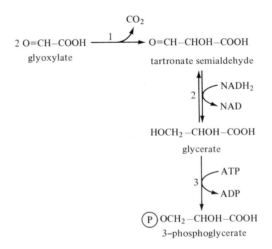

Figure 6.8. The glycerate pathway. 1, glyoxylate carboligase; 2, tartronate semialdehyde reductase; 3, glycerate kinase.

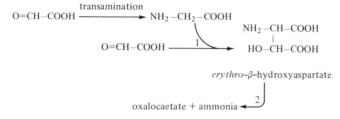

Figure 6.9. The β-hydroxyaspartate pathway. 1, *erythro-β-hydroxyaspartate* aldolase; 2, *erythro-β-hydroxyaspartate* dehydratase.

The metabolism of glyoxylate in *Paracoccus denitrificans* is different. Glycine is formed from glyoxylate by transamination. It then condenses with another molecule of glyoxylate to form *erythro-β-hydroxyaspartate*, which is converted to oxaloacetate and ammonia (Figure 6.9). **Glycollate** is metabolized by *P. denitrificans* by a similar route.

The most highly oxidized C_2-compound is **oxalate**, and it is rather surprising that some microorganisms are able to utilize it as carbon and energy source. One of these is *Pseudomonas oxalaticus*, which was studied in detail by Quayle. It gains reducing power by the oxidation of oxalate to CO_2 via oxalyl-CoA and formate (Figure 6.10).

Synthesis of cell constituents begins with the reduction of oxalyl-CoA to glyoxylate:

$$\text{oxalyl-CoA} + \text{NADPH}_2 \longrightarrow \text{glyoxylate} + \text{NADP} + \text{CoA}$$

Glyoxylate is then metabolized further via the glycerate pathway.

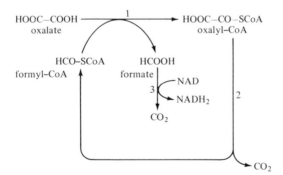

Figure 6.10. Oxidation of oxalate by *P. oxalaticus*. 1, oxalate is converted to oxalyl-CoA in a CoA transferase reaction using formyl-CoA as donor; 2, oxalyl-CoA is decarboxylated to yield formyl-CoA, which is converted to formate by the transferase reaction; 3, formate is oxidized by formate dehydrogenase.

IV.   Growth with Aliphatic Hydrocarbons

Because of the inertness of hydrocarbons it is interesting that many micro-organisms are able to utilize these compounds for growth. The primary attack on hydrocarbons requires oxygen, so that growth on them is an obligately aerobic process. From the nature of the substrates utilized two groups of organisms can be envisaged.

(a) Bacteria growing with methane. Since growth on a C_1-compound requires special pathways, these bacteria deserve special attention and will be discussed in a later section of this chapter. Methane-oxidizing bacteria normally do not grow with other hydrocarbons.

(b) Organisms growing on hydrocarbons other than methane. This is a rather heterogeneous group. Relatively few bacterial species (mycobacteria, flavobacteria, *Nocardia*), are able to grow on ethane, propane, butane, and hydrocarbons up to C_8. However, utilization of long-chain hydrocarbons is widespread among microorganisms, and *n*-alkanes with 10 to 18 carbons are utilized with the greatest frequency and rapidity. Table 6.1 shows a small selection of organisms able to grow on long-chain hydrocarbons. It is noteworthy that yeasts and fungi also utilize such substrates. *E. coli*, *Entero-bacter aerogenes*, and *Bacillus subtilis* are unable to grow with hydrocarbons.

Hydrocarbons are water-insoluble compounds and their uptake is probably a difficult task. Electron microscopic investigations have shown that microorganisms growing with hydrocarbons accumulate these compounds in considerable quantities as intracytoplasmic inclusions. Therefore, bacteria, yeasts, and other fungi must have a very effective transport system for hydrocarbons. Part of this transport system is made up of glycolipids, which are located in the outer portion of the cell wall and bring about microemulsions of the hydrocarbons so that they may move across the cell membrane.

In most cases the breakdown of a hydrocarbon is initiated by the oxidation of a terminal methyl group to a primary alcohol group. In *Pseudomonas*

Table 6.1. Organisms capable of growing with long-chain hydrocarbons

---

Pseudomonas fluorescens
P. aeruginosa
Acinetobacter calcoaceticus
Mycobacterium smegmatis
Nocardia petroleophila
Candida lipolytica
Torulopsis colliculosa
Cephalosporium roseum
Arthrobacter paraffineus
Arthrobacter simplex
Corynebacterium glutamicum

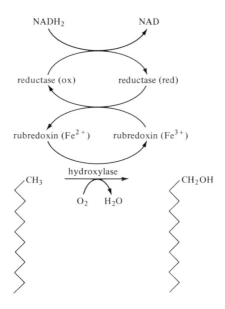

Figure 6.11. Terminal oxidation of an *n*-alkane by *Pseudomonas oleovorans*.

oleovorans this reaction requires three proteins, rubredoxin (which is an iron-sulfur protein), $NADH_2$-rubredoxin reductase, and alkane 1-hydroxylase. As shown in Figure 6.11 rubredoxin is first reduced with $NADH_2$. The oxidation of the hydrocarbon is then catalyzed by a mono-oxygenase (or hydroxylase), which requires an electron donor (reduced rubredoxin) in addition to the substrate. Instead of rubredoxin a different hydrogen donor may function as co-substrate in other alkane-oxidizing organisms. The general equation for this type of reaction is:

$$\text{substrate-H} + O_2 + AH_2 \xrightarrow[\text{(hydroxylase)}]{\text{mono-oxygenase}} \text{substrate-OH} + H_2O + A$$

The primary alcohol formed from the hydrocarbon is oxidized first to the aldehyde and then to the corresponding fatty acid by NAD-dependent dehydrogenases. Finally, fatty acids are degraded by β-oxidation as has already been described.

It should be mentioned that subterminal attack of hydrocarbons by oxygen has also been reported. Furthermore, a diterminal attack of oxygen can occur; the fatty acid produced in the first round of reactions is again hydroxylated at the terminal methyl group so that a dicarboxylic acid is formed, which then is subject to β-oxidation.

V.   Growth with Aromatic Compounds

Large amounts of compounds containing aromatic rings are produced by plants; the most predominant of these is lignin. Aromatic amino acids and vitamins are constituents of every organism. All these compounds become available when organisms die and organic matter is decomposed; they are degraded by bacteria and fungi. Mammals can only degrade phenylalanine and tyrosine but not tryptophan and other aromatic compounds.

The pathway used by animals and bacteria for phenylalanine and tyrosine degradation is outlined in Figure 6.12. A key intermediate involved is **homogentisate**, which is oxidized to maleylacetoacetate and metabolized further to fumarate and acetoacetate. A number of bacteria are able to cleave **gentisate** in a similar fashion, with fumarate and pyruvate being formed. Gentisate can be formed in *Pseudomonas acidovorans*, for instance, by

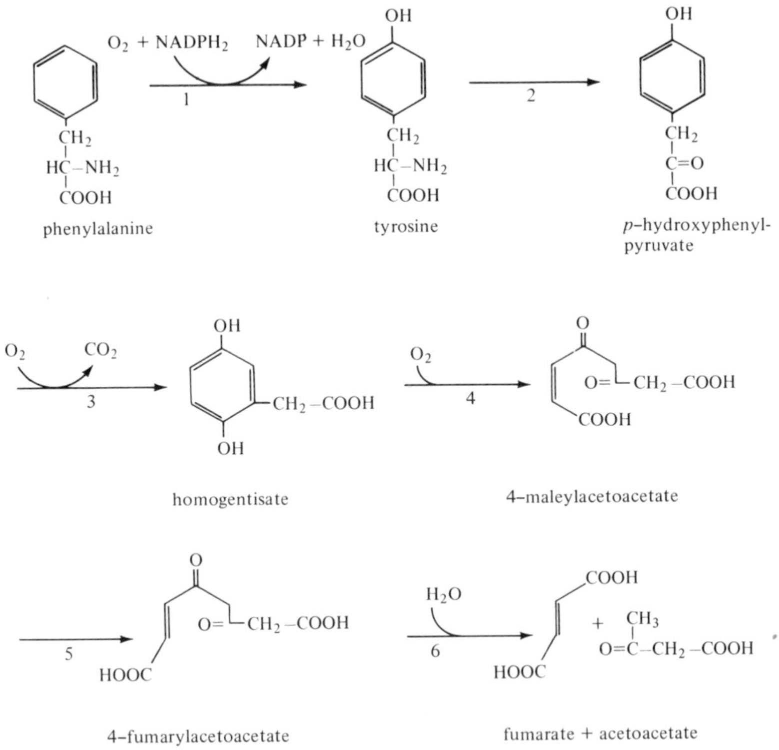

Figure 6.12. Degradation of phenylalanine and tyrosine by the homogentisate pathway. 1, phenylalanine hydroxylase; 2, transamination reaction; 3, *p*-hydroxyphenylpyruvate oxidase, which catalyzes decarboxylation, migration of the side chain, and hydroxylation of the ring; 4, homogentisate oxidase; 5, maleylacetoacetate isomerase (requires glutathione); 6, fumarylacetoacetate hydrolase.

hydroxylation of *m*-hydroxybenzoate.

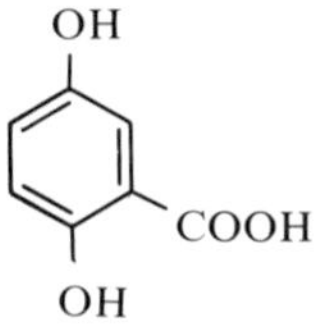

gentisate

Gentisate and homogentisate, however, are not the only compounds that can be oxidatively cleaved to nonaromatic products. In fact, the majority of aromatic compounds is converted by bacteria into **catechol** and **protocatechuate**, which—as discovered by Stanier and collaborators—are the "starting substrates" in the subsequent oxidative cleavage reactions. Figures 6.13 and 6.14 show catabolic routes leading to catechol and protocatechuate formation. It is apparent that pathways exist by which compounds such as

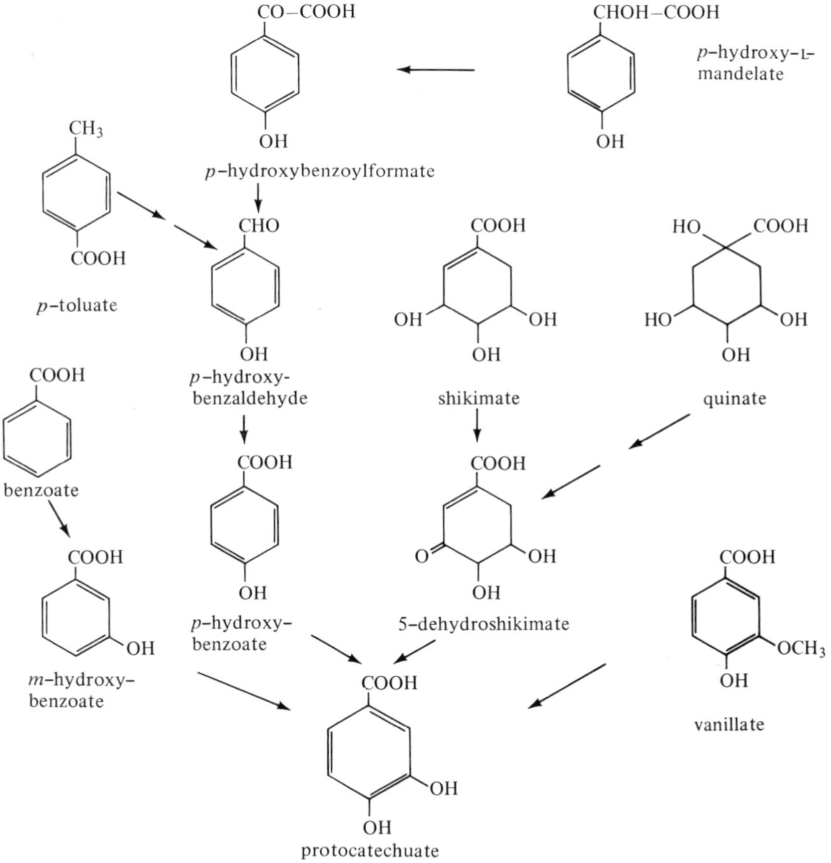

Figure 6.13. Aromatic and hydro-aromatic compounds that can be converted to protocatechuate. [R. Y. Stanier and L. N. Ornston, *Adv. Microbial Physiol.* **9**, 89–151 (1973).]

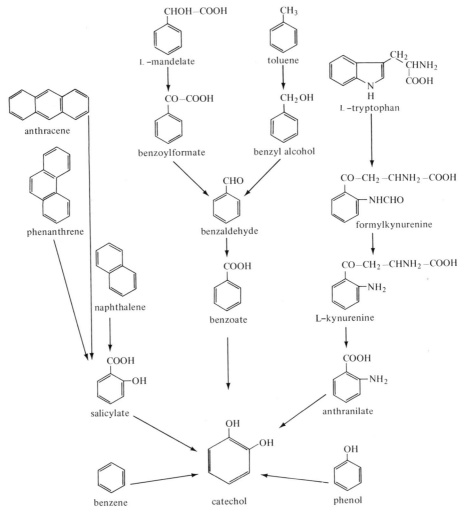

Figure 6.14. Aromatic compounds that can be converted to catechol. [R. Y. Stanier and L. N. Ornston, *Adv. Microbial Physiol.* **9**, 89–151 (1973).]

anthracene, mandelate, tryptophan, or quinate can be metabolized to the "starting substrates" mentioned. In fact, Figures 6.13 and 6.14 show only a small selection of compounds.

Two types of cleavage reactions are known for catechol and protocatechuate: ortho-cleavage and meta-cleavage.

A.   Ortho-cleavage or 3-oxoadipate pathway

As illustrated in Figure 6.15 the aromatic rings of catechol and protocate-chuate are cleaved, via oxygenase reactions, between the two hydroxyl groups. The products—*cis,cis*-muconate and 3-carboxy-*cis,cis*-muconate— yield then, in two reactions, the first common intermediate of these path-ways—4-oxoadipate enol-lactone. This compound is degraded further to yield succinate and acetyl-CoA. The reactions of the catechol and the protocatechuate branch are catalyzed by different sets of enzymes; there is,

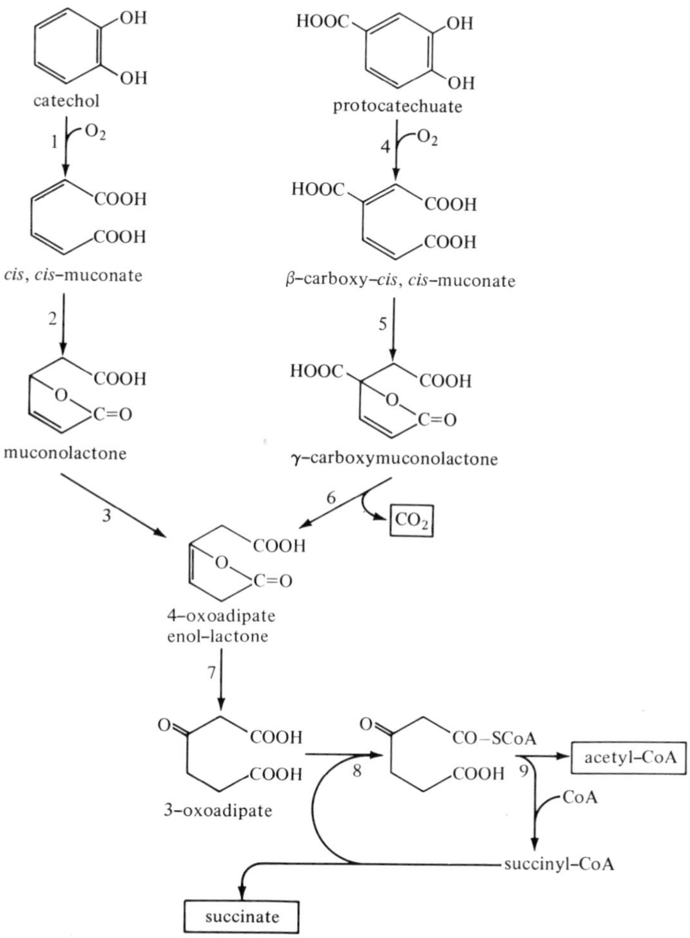

Figure 6.15. Reactions of the 3-oxoadipate pathway. 1, catechol 1,2-oxygenase; 2, muconate-lactonizing enzyme; 3, muconolactone isomerase; 4, protocatechuate 3,4-oxygenase; 5, β-carboxymuconate-lactonizing enzyme: 6, γ-carboxymuconolactone decarboxylase; 7,4-oxoadipate enol-lactone hydrolase; 8,3-oxoadipate succinyl-CoA transferase; 9, 3-oxoadipyl-CoA thiolase.

for instance, a catechol-1,2-oxygenase and a protocatechuate-4,5-oxygenase. This type of reaction can be illustrated as follows:

catechol-1, 2-oxygenase reaction

In contrast to the hydroxylase, or mono-oxygenase (see oxidation of hydrocarbons), an additional hydrogen donor such as $NADPH_2$ is not required in the dioxygenase reaction; both oxygen atoms appear in the product.

B.   Meta-cleavage

In 1959 Dagley and Stopher found that a soil pseudomonad contained enzyme systems that catalyze the breakdown of catechol and protocatechuate in a different way. The ring is opened adjacent to the hydroxyl groups, so that 2-hydroxymuconic semialdehyde or 2-hydroxy-4-carboxymuconic semi-aldehyde is formed. Their further metabolism leads to the formation of pyruvate, formate, and acetaldehyde (Figure 6.16).

The distribution of the various pathways among the bacteria is very complex. *Pseudomonas acidovorans* and *P. testosteroni* metabolize proto-catechuate-yielding substrates via the meta-cleavage pathway and catechol-yielding substrates via the ortho-cleavage pathway. In other pseudomonads the ortho-cleavage pathway dominates; this might be true for many bacteria that grow with aromatic substrates.

VI.   Growth with C_1 - Compounds

In connection with the energy metabolism of aerobes it was mentioned that a few groups of bacteria cannot employ the tricarboxylic acid cycle for the production of reducing power. One of these groups comprises the micro-organisms growing on C_1-compounds. Two subgroups can be envisaged:

1. Obligate methylotrophs, which grow only at the expense of compounds containing no carbon–carbon bonds (methane, methanol, etc.).
2. Facultative methylotrophs, which grow on a variety of carbon sources including C_1-compounds. Most of these organisms utilize methanol and methylamine but not methane.

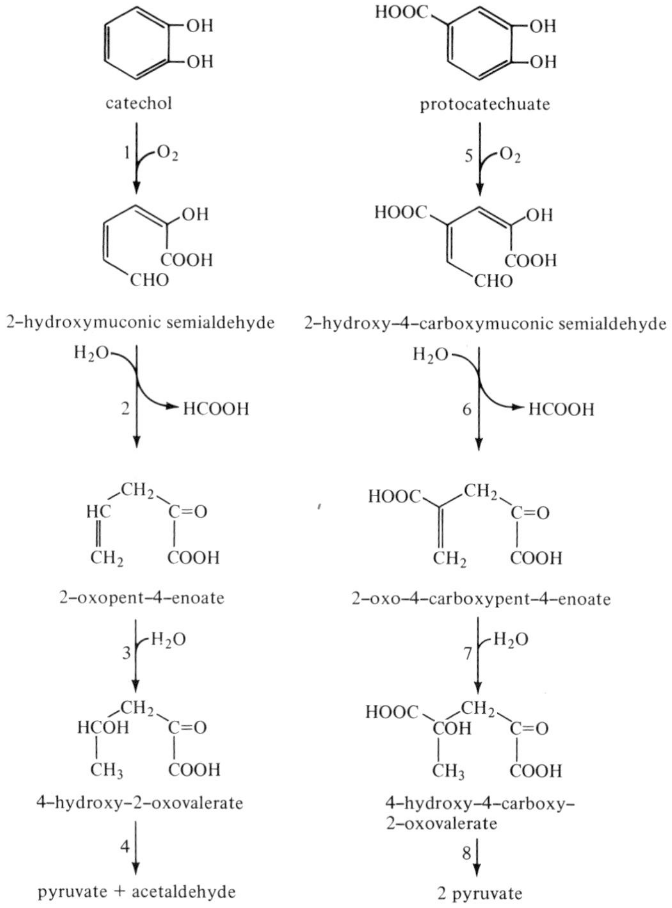

Figure 6.16. Dissimilation of catechol and protocatechuate by the pathways involving meta-cleavage. 1, catechol 2,3-oxygenase; 2, 2-hydroxymuconic semialdehyde hydrolase; 3, 2-oxopent-4-enoic acid hydrolase; 4, 4-hydroxy-2-oxovalerate aldolase; 5, protocatechuate 4,5-oxygenase; 6, 2-hydroxy-4-carboxymuconic semialdehyde hydrolase; 7, 2-oxo-4-carboxypent-4-enoic acid hydrolase; 8, 4-hydroxy-4-carboxy-2-oxovalerate aldolase.

A. Obligate methylotrophs

The first methylotroph isolated was *Bacillus methanicus* (Söhngen, 1906). Fifty years later it was reisolated and named *Pseudomonas methanica*. During the last 10 years a large number of methylotrophic bacteria became known through the work of Whittenbury and Wilkinson; on the basis of their shape, their internal membrane structure, and their ability to form cysts,

$$CH_4 + O_2 + NADH_2 \xrightarrow{1} CH_3OH + H_2O + NAD$$

$$CH_3OH + X \xrightarrow{2} CH_2O + XH_2$$

$$CH_2O + X + H_2O \xrightarrow{3} HCOOH + XH_2$$

$$HCOOH + NAD \xrightarrow{4} CO_2 + NADH_2$$

Figure 6.17. Oxidation of methane to CO_2. [T. Ferenci, T. Strøm, and J. R. Quayle, *J. Gen. Microbiol.* **91**, 79–91 (1975).] 1, mono-oxygenase reaction; 2, methanol dehydrogenase; 3, formaldehyde dehydrogenase; 4, formate dehydrogenase. X is an unknown electron carrier.

they are distinguished as *Methylosinus*, *Methylocystis*, *Methylomonas*, *Methylobacter*, and *Methylococcus*.

Clearly, during growth on methane, reducing power for the respiratory chain can only be produced by oxidation of methane to CO_2. There is no acetyl-CoA available to be oxidized via the tricarboxylic acid cycle. Oxidation of methane proceeds as shown in Figure 6.17. Methane is first oxidized by a mixed function oxidase (mono-oxygenase) to methanol. The oxidation of methanol to formaldehyde and of formaldehyde to formate is coupled to the reduction of the carrier X to XH_2; the nature of this electron carrier is not known. Finally, in the formate dehydrogenase reaction $NADH_2$ is generated for the initial step of methane oxidation.

It should be mentioned that one experimental result cannot be explained on the basis of the above scheme, namely that the amount of cells obtained from growth on 1 mol of methane is greater than the amount obtained with 1 mol of methanol. If methanol serves as substrate, 2 XH_2 are available as hydrogen donor and in addition one $NADH_2$. This should result in the production of more ATP and hence of more cells as compared with methane as substrate. A reasonable explanation cannot be given at the moment. However, it is possible that cells growing on methanol have to invest more ATP in substrate transport.

The anabolic metabolism of the methylotrophs diverges from the energy metabolism at the level of formaldehyde. Two formaldehyde fixation cycles have been elucidated, largely due to work done in the laboratories of Quayle and Hersh.

1. The serine-isocitrate lyase pathway. The formation of acetyl-CoA from formaldehyde and CO_2 in the serine pathway is shown in Figure 6.18. The acceptor for formaldehyde is glycine, and the serine formed in the hydroxymethylase reaction is converted to the corresponding α-oxo acid via a transaminase reaction. Hydroxypyruvate is then further converted to phospho-enolpyruvate by the action of several enzymes. Oxaloacetate is formed through the PEP carboxylase reaction. The key enzyme for the generation of glyoxylate—malyl-CoA lyase—came under study only recently, despite the fact that this cleavage reaction was discovered 15 years ago in *Rhodopseudomonas sphaeroides*.

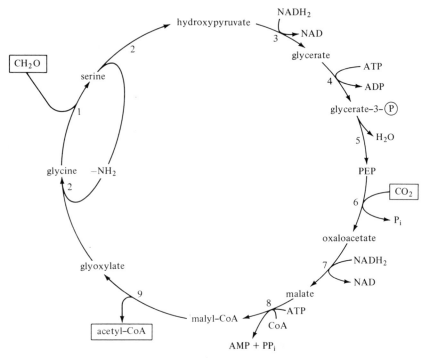

Figure 6.18. Formation of acetyl-CoA from formaldehyde and CO_2 in the serine pathway. 1, serine hydroxymethylase; 2, a transaminase that converts serine into hydroxypyruvate and glyoxylate into glycine; 3, hydroxypyruvate reductase; 4, glycerate kinase; 5, phosphoglycerate mutase and enolase; 6, PEP carboxylase; 7, malate dehydrogenase; 8, malyl-CoA synthetase; 9, malyl-CoA lyase.

The enzyme system (malyl-CoA synthetase) responsible for the formation of malyl-CoA from malate requires, of course, ATP. Thus far it has not been possible to detect it in all of those microorganisms that contain malyl-CoA lyase. Besides malyl-CoA synthetase and lyase two other enzymes can be considered as key enzymes of this cycle, serine-glyoxylate transaminase and hydroxypyruvate reductase.

The result of the serine pathway is the formation of acetyl-CoA from CH_2O and CO_2. The cycle is insufficient to supply the cell with PEP or oxaloacetate for biosynthetic purposes. This, however, is achieved when it is

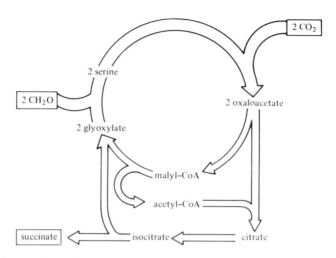

Figure 6.19. Net formation of succinate from formaldehyde and CO_2 by the serine-isocitrate lyase pathway.

combined with the citrate synthase, *cis*-aconitase, and isocitrate lyase reactions (Figure 6.19). In essence, this accomplishes the net formation of succinate from 2 CO_2 and 2 formaldehyde. The serine-isocitrate lyase pathway is present in *Methylosinus* and *Methylocystis* and a number of facultative C_1-utilizers.

2. The ribulose-monophosphate cycle. The key reactions of this cycle are the condensation of ribulose-5-phosphate and formaldehyde by hexulose-6-phosphate synthase and the isomerization of the product to fructose-6-phosphate.

$$
\begin{array}{ccccc}
\text{CH}_2\text{OH} & & \text{CH}_2\text{OH} & & \text{CH}_2\text{OH} \\
| & & | & & | \\
\text{C}=\text{O} & & \text{HOCH} & & \text{C}=\text{O} \\
| & & | & & | \\
\text{HCOH} & + \text{CH}_2\text{O} \longrightarrow & \text{C}=\text{O} & \longrightarrow & \text{HOCH} \\
| & & | & & | \\
\text{HCOH} & & \text{HCOH} & & \text{HCOH} \\
| & & | & & | \\
\text{CH}_2\text{O}\textcircled{P} & & \text{HCOH} & & \text{HCOH} \\
\text{ribulose-5-}\textcircled{P} & & | & & | \\
& & \text{CH}_2\text{O}\textcircled{P} & & \text{CH}_2\text{O}\textcircled{P} \\
& & \text{D-}\textit{erythro-}\text{1-glycero-3-} & & \text{fructose-6-}\textcircled{P} \\
& & \text{hexulose-6-}\textcircled{P} &
\end{array}
$$

The acceptor—ribulose-5-phosphate—is regenerated through the action of transketolase and transaldolase as shown in Figure 6.20. Three molecules of formaldehyde finally end up in one molecule of dihydroxyacetonephosphate, which can be utilized for biosynthetic purposes. The ribulose-monophosphate cycle occurs in *Methylococcus* and *Methylomonas* species.

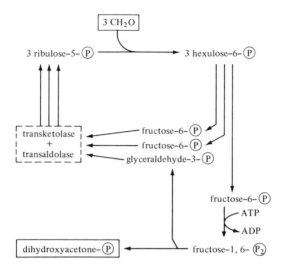

Figure 6.20. The ribulose-monophosphate cycle.

B. Facultative methylotrophs

A number of microoganisms besides the obligate methylotrophs are able to grow with C_1-compounds and more complex organic compounds. Among these are yeasts, *Hyphomicrobium* species, pseudomonads such as *P. oxalaticus* and *P.* AM1, and *Protaminobacter* species. These organisms utilize methanol, formate, or methylamine, but are unable to grow with methane. Recently, however, a bacterium (*Methylobacterium organophilum*) has been isolated, which is a facultative methylotroph, but also possesses the ability to utilize methane.

When growing with methanol, facultative methylotrophs employ either the serine-isocitrate lyase pathway or the ribulose-monophosphate cycle. Formate is assimilated by *P. oxalaticus* via CO_2 and the ribulose-1,5-bisphosphate cycle (Chapter 9).

VII. Incomplete Oxidations

During growth of aerobic heterotrophs, a part of the substrate is normally oxidized to CO_2 and the remainder is used to synthesize cellular material. In some organisms a complete oxidation of organic substrate is not possible and partially oxidized compounds—usually organic acids—are formed and excreted. The suboxydans group of the acetic acid bacteria, for instance, is unable to oxidize acetate. These bacteria grow with ethanol, and their metabolism may be described as follows.

Ethanol is oxidized with alcohol and acetaldehyde dehydrogenases to yield acetate and 2NADH$_2$:

$$CH_3CH_2OH + NAD \underset{}{\overset{\text{alcohol}}{\underset{\text{dehydrogenase}}{\rightleftharpoons}}} CH_3-C\overset{O}{\underset{H}{\diagdown}} + NADH_2$$

$$CH_3-C\overset{O}{\underset{H}{\diagdown}} + NAD + H_2O \xrightarrow{\substack{\text{aldehyde} \\ \text{dehydrogenase}}} CH_3-COOH + NADH_2$$

The NADH$_2$ thus produced is used for the production of ATP in the respiratory chain; acetyl-CoA cannot be oxidized via the tricarboxylic acid cycle and, therefore, most of the acetate formed is excreted.

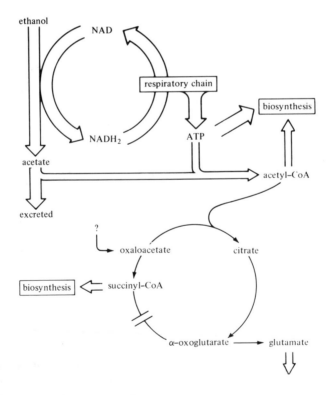

Figure 6.21. Metabolic situation of *Gluconobacter suboxydans*. Some reactions of the TCC serve as anabolic sequence; the cycle as a whole is interrupted. Reactions leading to oxaloacetate generation are not known.

The metabolic situation of *Gluconobacter suboxydans* is illustrated in Figure 6.21. It is apparent that the inability to oxidize acetyl-CoA is not due to the lack of the tricarboxylic acid cycle *per se*. In fact, citrate synthase, cis-aconitase, and isocitrate dehydrogenase are present and employed for glutamate synthesis; α-oxoglutarate dehydrogenase, however, is absent. The suboxydans group lacks also isocitrate lyase and malate synthase and the nature of the anaplerotic sequence required for PEP and oxaloacetate formation is as yet unknown.

Other acetic acid bacteria (*Acetobacter aceti, A. xylinum, A. peroxydans*) differ from *G. suboxydans* in that after exhaustion of ethanol the acetate excreted is taken up again and oxidized to CO_2. The rate of this acetate oxidation is low in *A. aceti* and *A. xylinum* and moderate in *A. peroxydans*. These species possess a functional tricarboxylic acid cycle.

In addition to ethanol the acetic acid bacteria oxidize a large number of other alcohols to the corresponding acids and ketones. Examples are:

$$\text{propanol} \longrightarrow \text{propionate}$$
$$\text{isopropanol} \longrightarrow \text{acetone}$$
$$\text{glycerol} \longrightarrow \text{dihydroxyacetone}$$
$$\text{glucose} \longrightarrow \text{gluconate}$$
$$\text{gluconate} \longrightarrow \text{5-ketogluconate}$$

Again, these organisms are unable to synthesize the catabolic enzymes for the degradation of these substrates in high activity and excrete acids and ketones. Of special interest is the oxidation of D-sorbitol by acetic bacteria to L-sorbose. The latter is required in large amounts for the synthesis of vitamin C (Figure 6.22).

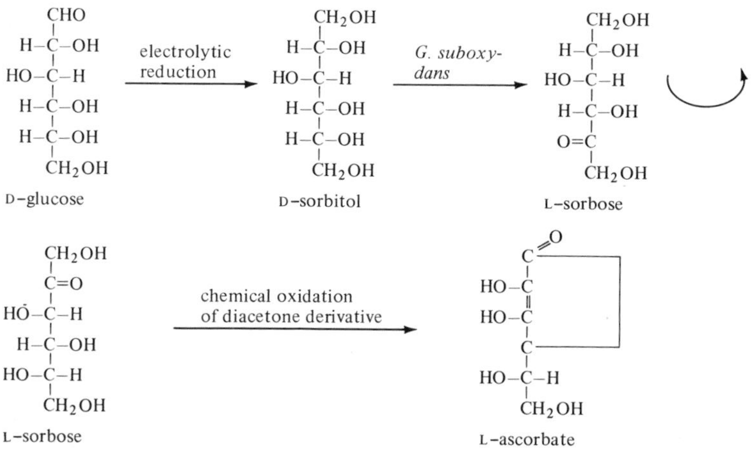

Figure 6.22. Synthesis of vitamin C. Electrolytic reduction of D-glucose yields D-sorbitol, which is oxidized by *G. suboxydans* to L-sorbose. Chemical oxidation leads to formation of the vitamin.

Figure 6.23. Incomplete oxidation of glucose during aerobic growth of *B. megaterium*. (a) About 30% of the glucose utilized is converted into acetate, pyruvate, and 2,3-butanediol. (b) Following depletion of glucose the products of incomplete oxidation are oxidised to yield the energy required for sporulation.

A number of microorganisms change from complete oxidation to incomplete oxidation under abnormal physiological conditions (high substrate concentrations or extreme pH). Yeasts and other fungi excrete citrate, fumarate, and gluconate under certain conditions and are employed for the production of these acids on an industrial scale. *Corynebacterium glutamicum* can excrete large amounts of L-glutamate. This microorganism and other bacteria (as well as mutants) are used at the present time to produce amino acids and other metabolites.

Most bacilli carry out incomplete oxidations when growing on carbohydrates. As shown in Figure 6.23 glucose is partly converted by *B. megaterium* into acetate, pyruvate, acetoin, and 2,3-butanediol during aerobic growth. These compounds are taken up again when glucose is depleted. Then they are oxidized to CO_2 and thus serve as source of ATP for the sporulation process. Acetoin and 2,3-butanediol are formed from pyruvate via α-acetolactate. During sporulation acetate is formed from these C_4-compounds by the **2,3-butanediol cycle** (Figure 6.24). In this cycle diacetyl is cleaved into acetate and enzyme-bound hydroxyethyl-thiamin pyrophosphate. The hydroxyethyl residue is transferred to another molecule of diacetyl to yield diacetylmethylcarbinol. The latter is reduced and cleaved into acetate and butanediol, which subsequently is oxidized to diacetyl. The acetate thus produced is fed into the tricarboxylic acid cycle.

There is also evidence for a direct cleavage of acetoin to acetate and acetaldehyde:

$$CH_3-\underset{\underset{O}{\|}}{C}-\underset{\underset{OH}{|}}{CH}-CH_3 \quad \xrightarrow[\;X\quad\quad XH_2\;]{H_2O} \quad CH_3-COOH + CH_3-CHO$$

The natural electron acceptor (X) is not known; *in vitro* dichlorophenol-indophenol can be used. The cleavage enzyme is identical with diacetyl-methylcarbinol synthase, and it is possible that acetoin cleavage is more important in acetate formation from these C_4-compounds than is the 2,3-butanediol cycle.

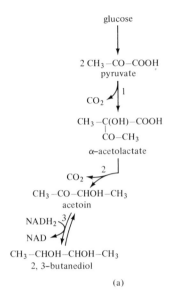

(a)

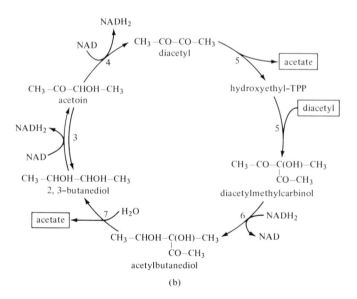

(b)

Figure 6.24. Formation of acetoin and 2,3-butanediol during growth of bacilli on glucose (A) and acetate formation by the 2,3-butanediol cycle during sporulation (B). 1, α-acetolactate synthase, a thiamin pyrophosphate-containing enzyme; 2, α-acetolactate decarboxylase; 3, 2,3-butanediol dehydrogenase; 4, acetoin dehydrogenase; 5, diacetylmethylcarbinol synthase; 6, diacetylmethylcarbinol reductase; 7, acetylbutanediol hydrolase.

VIII.   Summary

1. The polymers of dead organic material are hydrolyzed by exoenzymes, which are excreted by bacteria and fungi. α-Amylase cleaves starch into dextrins and subsequently into maltose, glucose, and oligosaccharides. Cellulase hydrolyzes cellulose to cellobiose. The disaccharides are taken up by bacteria and cleaved to hexose-1-phosphate and hexose by specific phosphorylases.

2. Proteases are formed by bacilli, pseudomonads, and many anaerobes; DNases are excreted by hemolytic streptococci, *Staphylococcus aureus*, and clostridia. *Bacillus subtilis* excretes a powerful RNase.

3. The degradation of many amino acids is initiated by their conversion to the corresponding α-oxo acids. This is accomplished either by cytochrome-linked oxidases or by NAD(P)-linked dehydrogenases and transaminases. Serine and threonine are converted into α-oxo acids by dehydratases. Many bacteria form aspartase, which removes ammonia from aspartate to yield fumarate.

4. Fatty acids are degraded by β-oxidation. Bacteria growing with these compounds employ the glyoxylate cycle for the formation of C_4-dicarboxylic acids.

Propionate is utilized either by the methylmalonyl-CoA pathway or by the acrylyl-CoA pathway. Glyoxylate is a degradation product of the purine bases; it is oxidized via the dicarboxylic acid cycle in which malate — formed from glyoxylate and acetyl-CoA—is oxidized to acetyl-CoA. PEP for biosynthesis is formed via tartronate semialdehyde by the glycerate pathway. *Pseudomonas oxalaticus* grows with oxalate, which is oxidized to CO_2 via formate for ATP formation and reduced to glyoxylate for synthesis of cell constituents.

5. Aliphatic hydrocarbons are oxidized at one terminal methyl group by mono-oxygenases to the corresponding alcohols. The alcohols are further oxidized to fatty acids by NAD-dependent dehydrogenases. Subterminal and diterminal attack of hydrocarbons by oxygen has also been reported.

6. The majority of aromatic compounds utilized by bacteria is first converted into catechol and protocatechuate. Two kinds of cleavage reactions for these compounds occur in bacteria: (1) ortho-cleavage, in which catechol is oxidized to *cis, cis*-muconic acid; (2) meta-cleavage, in which catechol is oxidized to 2-hydroxymuconic semialdehyde.

7. Methylotrophs utilize methane and other C_1-compounds for growth. Reducing power for the respiratory chain is formed by these organisms by the oxidation of the C_1-compound to CO_2. Formaldehyde is the starting material for biosynthesis. It is assimilated either by the ribulose-monophosphate cycle or by the serine-isocitrate lyase pathway.

8. Acetic acid bacteria gain $NADH_2$ for the respiratory chain by the oxidation of ethanol to acetate. The latter is excreted because of the incompleteness or low capacity of the tricarboxylic acid cycle. During growth,

bacilli oxidize sugars to acetate, pyruvate, and 2,3-butanediol. During the sporulation process these compounds serve as an energy source and are oxidized to CO_2.

Chapter 7
Regulation of Bacterial Metabolism

The discussion of the metabolism of aerobic heterotrophs has shown that a number of catabolic and anaplerotic enzymes are needed by microorganisms only under certain growth conditions. β-Galactosidase is required by *E. coli* during growth on lactose but not if glucose serves as substrate. A pseudomonas confronted with phenol in its environment requires several specific enzymes in order to take advantage of this compound. These enzymes are not needed if organic acids such as malate or succinate were the substrates utilized by this microorganism. Thus, it is reasonable and economical that organisms do not synthesize all the time all the enzymes they are able to make but only those that are necessary for their metabolism under current physiological conditions. This regulation of enzyme synthesis is accomplished by induction and repression.

Alternatively, the activity of the enzymes present in the cell has to be under control. Enzyme-catalyzed reactions have to proceed in accordance with the demands of the cell for energy and for cellular constituents. The cell must, therefore, contain devices to slow down or to speed up the synthesis of a particular amino acid or the formation of ATP. This control is accomplished by the ability of the cell to increase or to decrease the activity of certain key enzymes of the metabolic pathways.

I. Regulation of Enzyme Synthesis by Induction and Repression

A. Enzyme induction

Figure 7.1 shows the increase of the level of β-galactosidase in *E. coli* cells following the addition of lactose to a culture medium. In the absence of lactose the level of this enzyme in *E. coli* is so low that it is detectable only with special techniques. Lactose increases it by a factor of approximately 1,000. As early as 1.4 minutes after lactose addition an increase of enzyme activity is already measurable. Enzyme activity reaches a plateau after 15 to 180 minutes depending on the growth conditions. This process of substrate-mediated enzyme synthesis is called enzyme induction. Actually allolactose (α-D-galactosyl-β-1,6-D-glucose), a compound derived from lactose (α-D-galactosyl-β-1,4-D-glucose), functions as inducer.

We know already that more than one enzmye is usually required to channel a substrate into the intermediary metabolism. For the breakdown of lactose the permease, β-galactosidase, and the enzymes that convert galactose into glucose-1-phosphate are necessary (see Chapter 4). For growth on an aromatic compound *Pseudomonas putida* has to synthesize about 10 enzymes that are not present in cells growing on other substrates. What is the inducer for all these enzymes? Two kinds of mechanisms can be distinguished, **coordinated** and **sequential induction**. In the first case, the growth substrate switches on the synthesis of all enzymes required for its degradation. This is found for short catabolic sequences. The enzymes of the Entner–Doudoroff pathway are formed by coordinated induction, also the enzymes for the metabolism of arabinose. Lactose induces a specific permease, β-galactosidase, and a transacetylase. Galactose formed in the β-galactosidase reaction induces the enzymes for its conversion into glucose-

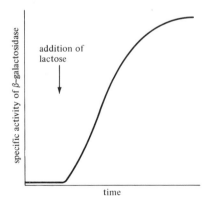

Figure 7.1. Increase of the specific activity of β-galactosidase in *E. coli* following the addition of lactose to a cell suspension.

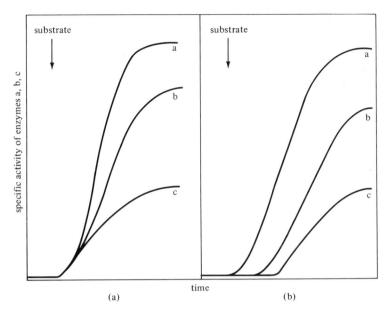

Figure 7.2. Coordinated (a) and sequential (b) induction of enzymes a, b, and c.

1-phosphate. Coordinated induction is indicated when upon addition of the substrate all enzymes are induced almost simultaneously, as shown in Figure 7.2.

Sequential induction is found for long catabolic pathways that serve for the degradation of several substrates.

If the substrates A, B, C, and D are catabolized as indicated, a few enzymes are necessary for the breakdown of all four substrates, some are required for two substrates, and some for only one, A, B, or C. Here sequential induction lends itself as an economical principle. One of the best studied examples of sequential induction is the formation of the enzymes for the degradation of aromatic compounds. The inducers for the synthesis of the enzymes of the 3-oxoadipate pathway in *Pseudomonas putida* are given in Figure 7.3. It is obvious that the two branches of this pathway are regulated in a different manner. Protocatechuate is the inducer of protocatechuate oxygenase. The four subsequent enzymes are coordinatedly induced by one of the terminal intermediates, 3-oxoadipate. In the catechol branch, *cis,cis*-muconate acts as **product-inducer** of catechol 1,2-oxygenase and as **substrate-inducer** of muconate-lactonizing enzyme and muconolactone isomerase. The fact that catechol 1,2-oxygenase and the terminal enzymes of this pathway are

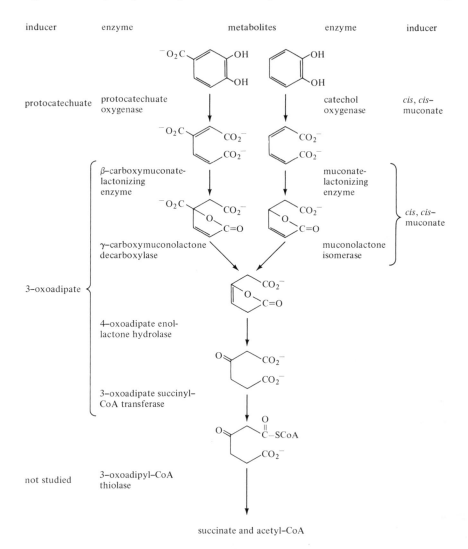

Figure 7.3. The inducers for the enzymes of the 3-oxoadipate pathway in *Pseudomonas putida*. [R. Y. Stanier and L. N. Ornston, *Adv. Microbial Physiol.* **9**, 89–151 (1973).]

product-induced requires that catechol be converted to *cis,cis*-muconate and muconolactone to 3-oxoadipate before the induction of the enzymes involved can be initiated. In uninduced cells low levels of the corresponding enzymes are indeed detectable (Table 7.1).

It should be noted that the kind of regulation of the 3-oxoadipate pathway as elucidated in *P. putida* cannot be generalized. The investigations of Stanier and collaborators have shown that species-specific differences are very large. In *Acinetobacter calcoaceticus*, all enzymes for the degradation of protocate-

Table 7.1. Levels if some enzymes of the 3-oxoadipate pathway in uninduced and induced cells of *Pseudomonas putida*[a]

enzyme	specific activity (U/mg protein)	
	uninduced (succinate)	induced (benzoate)
catechol 1,2-oxygenase	<0.2	1,090
cis,cis-muconate-lactonizing enzyme	<0.2	390
muconolactone isomerase	<20	2,200
4-oxoadipate enol-lactone hydrolase	30	1,490

[a]R. Y. Stanier and L. N. Ornston, *Adv. Microbial Physiol.* **9**, 89–151 (1973).

chuate are induced by that compound; *cis,cis*-muconate induces all enzymes required if catechol is the substrate. Still another regulatory map has been found for *Alcaligenes eutrophus*.

How does the inducer tell the cell to synthesize the enzymes of a catabolic sequence? In Chapter 3 we have discussed the mechanism of protein synthesis. That glucose-grown *E. coli* cells do not contain the enzymes for the catabolism of lactose and many other substrates means that the entire bacterial genome is not blindly transcribed into RNA sequences and further translated into proteins. Gene expression is controlled by sophisticated mechanisms. Several genes are always expressed in growing cells, for bacteria contain a number of enzymes that have to be present under all conditions (constitutive enzymes). Other genes are switched on or off just as required in a particular physiological situation. For a number of inducible enzymes, it is now known how this is accomplished. The best studied example is the formation of the enzymes for lactose catabolism in *E. coli*.

The three genes which code for lactose permease, β-galactosidase, and the transacetylase are located on the *E. coli* chromosome adjacent to one another. Together with the promotor and the operator region they form a functional unit, the **lac operon**. This unit is not transcribed in the absence of lactose because a repressor blocks the operator region [Figure 7.4(a)]. When lactose enters an *E. coli* cell some allolactose is formed by the action of β-galactosidase, which is present in uninduced cells in very small amounts. Allolactose is bound by the repressor protein; and the latter is thereby modified in such a way that it can no longer bind to the operator region; transcription can proceed [Figure 7.4(b)].

Presumably a number of inducible enzyme systems are regulated in a similar way. Sequential induction is also understandable on the basis of the operon model. The structural genes of the enzymes of a branched catabolic pathway can be located on the chromosome in the form of two,

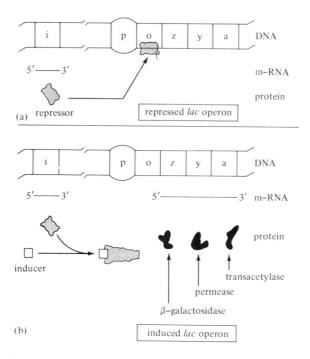

Figure 7.4. Induced synthesis of the enzymes for lactose catabolism. (a) The product of the gene i is a repressor protein that binds to the o region of the lac operon and thus prevents transcription of this operon. (b) The inducer (allolactose) binds to the repressor, which thereby loses its affinity to the o region, m-RNA is made, and the three proteins are formed. p, promotor; o, operator; z, y, and a, genes coding for β-galactosidase, permease, and transacetylase. Note that the lengths in base pairs of the structural genes and the p o region are very different and not identical as could be deduced from the presentation of the DNA in this figure.

three, or more functional units which are switched on or off by the appropriate inducers.

The *lac* operon is an example of a **negative control** mechanism. The operon itself is always ready to be transcribed, but transcription is prevented by a specific repressor. This repressor has to be removed by the inducer. In contrast, the operons for the enzymes of arabinose, maltose, and rhamnose catabolism—at least in *E. coli*—are under **positive control**. The operon has to combine with an activator—consisting of a specific protein and the inducer—in order to be transcribed.

B. Catabolite repression

In 1942 Monod discovered the phenomenon of **diauxie**. A culture of *Bacillus subtilis* supplemented with glucose and arabinose as carbon and energy sources utilized first glucose and then arabinose. This resulted in a biphasic

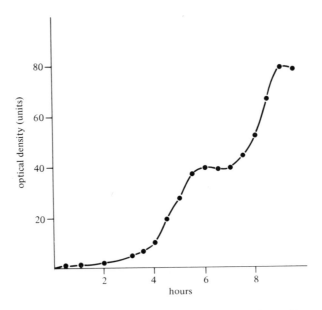

Figure 7.5. Diauxic growth of *B. subtilis* on glucose and arabinose. It was shown by additional experiments that glucose was utilized during the first period of growth and arabinose during the second. [J. Monod, *Recherches sur la croissance des cultures bactériennes* (thesis 1942), Hermann, Paris, p. 145 (1958).]

growth curve (Figure 7.5). The same growth behavior is observed for a great variety of substrate combinations utilized by *B. subtilis*, *E. coli*, and many other organisms. In all these cases one readily utilizable substrate represses the utilization of other substrates, and, therefore, Magasanik introduced the term catabolite repression for this phenomenon.

   E. coli, confronted with a mixture of glucose and lactose in its growth medium, will grow first at the expense of glucose. The three structural genes of the *lac* operon are not expressed during this period. Expression begins when the concentration of glucose has become low. How can glucose prevent lactose from initiating the formation of the lac enzymes? This question was the subject of intensive research work in many laboratories. Finally it was found that catabolite repression is connected to the level of cylic AMP in the cells.

   The utilization of glucose by *E. coli* leads to a tremendous decrease of the intracellular concentration of cyclic AMP. After consumption of most of

adenosine-3',5'-monophosphate
(cyclic AMP)

the glucose the level of this compound increases again, and it could be shown that cyclic AMP is required for the transcription of the *lac* operon. It forms a complex with the so-called **CRP protein (cyclic AMP receptor protein)** and this complex binds to the p region of the *lac* operon. Initiation of transcription of the *lac* operon is only possible if this complex is present. Thus the mechanism of enzyme induction as described in Figure 7.4(b) is incomplete; Figure 7.6(b) accounts for the recent discovery of the involvement of cyclic AMP. In accordance with this mechanism, catabolite repression of glucose toward lactose is overcome if high concentrations of cyclic AMP are added to the growth medium.

What makes the level of cyclic AMP decrease if glucose is being metabolized? The level of cyclic AMP in the cells is determined by the activity of the enzyme adenylate cyclase, which forms this compound from **ATP** according to the following equation:

$$\text{ATP} \xrightarrow{\text{adenylate cyclase}} \text{cyclic AMP} + \text{PP}_i$$

Adenylate cyclase is membrane-bound; its activity is high if components of the sugar transport (phosphotransferase system) are phosphorylated. This is the case in the absence of transportable sugars. In their presence, the degree of phosphorylation decreases because of the phosphorylation of the sugar molecules entering the membrane. Adenylate cyclase thus becomes less active; the level of cyclic AMP goes down, and the cell is under catabolite repression.

The fact that glucose itself does not cause catabolite repression, but the change in the concentration of a common metabolite, makes it understandable that this kind of repression is found for various substrate combinations. In *E. coli* glucose represses the induction of several catabolic pathways (for utilization of lactose, sorbitol, xylitol, arabinose, glycerol, etc.). Fructose and glucose-6-phosphate can replace glucose. As a rule it can be stated that substrates utilized rapidly by constitutive enzyme systems cause catabolite repression of inducible pathways. Glucose and fructose do not dominate in

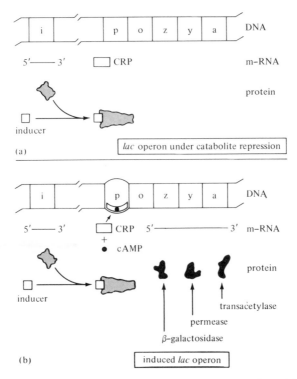

Figure 7.6. Induced synthesis of the enzymes for lactose catabolism. (a) In the absence of cyclic AMP, CRP is not bound at the promotor; m-RNA is not made. (b) The CRP–cyclic AMP complex is bound at the promotor; the structural genes z, y, and a are transcribed and the corresponding enzymes are synthesized.

all microorganisms. In *Clostridium tetanomorphum*, L-glutamate represses the formation of the enzymes for glucose utilization; in *Alcaligenes eutrophus*, a hydrogen-oxidizing chemolithotroph, H_2 prevents the induction of the Entner–Doudoroff pathway by fructose.

C.   End product repression

When *E. coli* grows in a minimal medium with glucose as carbon and energy source, all the monomers for the formation of macromolecules have to be synthesized along the pathways discussed in Chapter 3. Amino acids, nucleotides, etc., are needed in correct amounts for polymer synthesis, and an overproduction of some monomers has to be avoided. Sometimes certain monomers are available from the environment of the bacteria. If, for instance, histidine or tryptophan can be taken up, the biosynthesis of these compounds would be useless. Thus, for economical reasons bacteria must contain devices to regulate the level of anabolic enzymes. This requirement is met by the mechanism of repression of anabolic enzyme synthesis by the correspond-

ing end products. If tryptophan and histidine are added to a growing
E. coli culture, the cells stop the synthesis of the enzymes making histidine
from phosphoribosylpyrophosphate and tryptophan from chorismate.

For an understanding of end product repression the finding was important that the structural genes of several anabolic enzymes are also located
on the chromosome as operons. The nine genes responsible for the enzymes
of histidine biosynthesis in *Salmonella typhimurium* represent one operon.
The five genes coding for the enzymes that catalyze the synthesis of tryptophan from chorismate also form one operon [Figure 7.7(a)]. Repression
of enzyme synthesis is then understood as follows: when the product of a
pathway accumulates in the cells, it combines with a repressor protein to
give an active repressor. The latter binds to the operator region and prevents
transcription of the operon [Figure 7.7(b)].

End product repression stops enzyme synthesis within a short time
because of the short half-life of messenger RNA. The actual product does not
function as co-repressor in all anabolic sequences. In some amino acid
biosyntheses the tRNA derivative is the active agent, e.g., in histidine and
valine biosynthesis. It also should be noted that clustering of the genes of

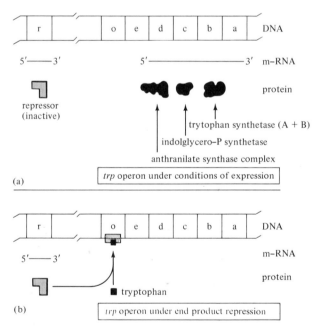

(a)

(b)

Figure 7.7. The tryptophan operon and the mechanism of end product repression.
(a) If tryptophan does not accumulate an inactive repressor is present; the structural
genes are transcribed and proteins are synthesized. The products of genes e and d form
the anthranilate synthase complex, gene c gives indolglycerophosphate synthetase
and a and b, the two components of tryptophan synthetase. (b) If tryptophan accumulates an active repressor is formed and transcription is repressed.

anabolic sequences on the chromosome is not a prerequisite for regulation by end product repression. In *E.coli*, the genes for the enzymes involved in arginine biosynthesis are located at various positions of the chromosome; they all are switched off by the same repressor–arginine complex.

Regulation of enzyme synthesis by end product repression becomes more difficult when we look at branched anabolic pathways. If here one end product would completely repress the formation of enzymes that serve also in the synthesis of other monomers, this could result in a shortage of these compounds and a stoppage of growth. An inspection of branched pathways with respect to the control of enzyme synthesis reveals that two principles are applied in order to avoid the above-mentioned difficulties (Figure 7.8):

1. **Isoenzymes** are formed for common reactions so that each end product has "its" enzyme whose synthesis can be repressed or derepressed. *E. coli* forms three aspartokinases and two homoserine dehydrogenases, which are under the control of different end products.
2. Repression of enzyme synthesis requires that all products synthesized by

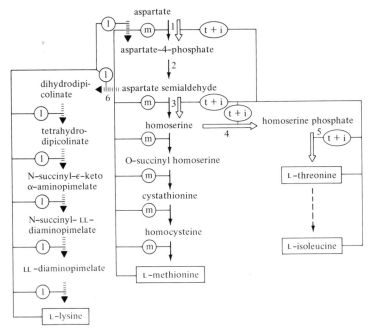

Figure 7.8. End product repression in the pathway leading from aspartate to the aspartate family of amino acids in *E. coli*. Enzymes indicated by hollow arrows are subject to divalent repression by L-threonine+L-isoleucine. L-Methionine represses the synthesis of the enzymes indicated by solid arrows and L-lysine those indicated by broken arrows. 1, aspartokinase; 2, aspartate semialdehyde dehydrogenase; 3, homoserine dehydrogenase; 4, homoserine kinase; 5, threonine synthase; 6, dihydrodipicolinate cyclohydrolase.

these enzymes are present in excess. This kind of repression is called **multivalent repression**, or di- or trivalent repression if two or three products are involved.

It is apparent from Figure 7.8 that aspartokinase I and homoserine dehydrogenase I (hollow arrows) are subject to divalent repression by L-threonine and L-isoleucine. This is reasonable. Since L-threonine is the precursor of L-isoleucine, end product repression only by L-threonine could bring the cell into difficulties. The level of aspartokinase II and of homoserine dehydrogenase II is controlled by L-methionine and that of aspartokinase III by L-lysine.

Another interesting branched pathway is the one for L-isoleucine, L-valine, and L-leucine synthesis. Here it is necessary to remember that each of the four steps involved in L-valine and L-isoleucine synthesis is catalyzed by the same enzyme and that α-oxo-β-methylbutyrate is not only the precursor of L-valine but also the starting material for L-leucine synthesis (Figure 7.9). In this case a trivalent repression mechanism is installed in order to prevent a shortage of one of the three amino acids. L-Leucine alone

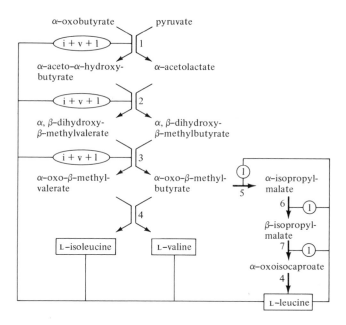

Figure 7.9. Regulation of synthesis of the enzymes involved in L-isoleucine, L-valine, and L-leucine formation in *Salmonella typhimurium*. i+v+1, trivalent repression by L-isoleucine + L-valine + L-leucine; 1, acetohydroxy acid synthase; 2, acetohydroxy acid isomero reductase; 3, dihydroxy acid dehydratase; 4, transaminase; 5, α-isopropylmalate synthase; 6, isopropylmalate isomerase; 7, β-isopropylmalate dehydrogenase.

represses the synthesis of the enzymes leading from α-oxo-β-methylbutyrate to L-leucine.

The synthesis of the enzymes involved in aromatic amino acid formation is also under control. *E. coli* contains three 7-phospho-3-deoxy-D-arabino-heptulosonate synthetases and two chorismate mutases. Regulation of enzyme synthesis by the end products phenylalanine, tyrosine, and tryptophan is quite similar to that encountered for the pathway of the aspartate family.

The regulatory maps discussed here have been elaborated for *E. coli* and *Salmonella typhimurium* and they cannot be generalized. In fact, comparative studies have shown that other solutions for the regulatory problems in branched pathways have been developed and are employed. *Bacillus polymyxa* and *Rhodospirillum rubrum* possess only one aspartokinase and one homoserine dehydrogenase; the synthesis of these enzymes is regulated by multivalent repression. *Bacillus subtilis* contains one 7-phospho-3-deoxy-D-arabinoheptulosonate synthetase but two chorismate mutases. It can be concluded that the anabolic pathways remained unchanged in the various microorganisms during evolution but that the mode of control of these pathways has been modified considerably.

D. Synthesis of enzymes of central pathways

The term induction is so much associated with catabolic enzymes and the term repression with anabolic enzymes that it is necessary to remember that the level of central enzymes also needs to be regulated. The requirement for anaplerotic sequences changes with the substrate. When *E. coli* is transferred from a glucose medium to an acetate medium, isocitrate lyase and malate synthase are formed. For faculatively anaerobic bacteria the change from aerobic to anaerobic growth has consequences with respect to the level of central enzymes. The complete tricarboxylic acid cycle is not needed under anaerobic conditions, and the bacteria stop to synthesize α-oxoglutarate

Table 7.2. Level of tricarboxylic acid cycle enzymes in *E. coli* grown aerobically or anaerobically with glucose[a]

enzyme	specific activity (U/mg protein)	
	aerobic growth	anaerobic growth
citrate synthase	51.1	10.5
cis-aconitase	317	16.1
isocitrate dehydrogenase	1,416	138
α-oxoglutarate dehydrogenase	17.4	0

[a]C. T. Gray, J. W. T. Wimpenny, and M. R. Mossman, *Biochim. Biophys. Acta* **117**, 33–41 (1966).

dehydrogenase. Citrate synthase, *cis*-aconitase, and isocitrate dehydrogenase are then required only for glutamate synthesis; their level in the cells is, therefore, lower than under aerobic conditions (Table 7.2). The mechanisms and metabolites that determine the level of enzymes of central pathways are not yet fully understood.

II.   Regulation of Enzyme Activity

A.   Feedback inhibition

In addition to the regulation of the level of enzymes the cell must have devices to adjust the activity of enzymes to the metabolic requirements. If a monomer is synthesized in larger amounts than is needed for polymer synthesis, it is not only necessary to stop the synthesis of the enzymes involved but also to reduce immediately the synthesis of that monomer. This is accomplished by **feedback inhibition**; the end product—when accumulating in the cell—inhibits the activity of the first enzyme involved specifically in its formation. If isoleucine is synthesized in excess, it inhibits the activity of the first enzyme of its biosynthetic pathway, threonine deaminase:

$$threonine \longrightarrow \alpha\text{-oxobutyrate} \to \to \to \to isoleucine$$

If pyrimidine nucleotides accumulate in the cell, CTP inhibits the activity of the first enzyme of their biosynthetic pathway, aspartate transcarbamylase:

$$\begin{array}{c} aspartate \\ + \\ carbamyl \\ phosphate \end{array} \longrightarrow carbamyl\ aspartate \to \to \to \to \to \to CTP$$

As soon as threonine deaminase and aspartate transcarbamylase are inhibited, the subsequent enzymes of these pathways run out of substrates, and monomer synthesis stops. It is now established that the substrate flow through most biosynthetic pathways is regulated by feedback inhibition. As it is in the regulation of the enzyme level, branched pathways are more difficult to be controlled in their activity than unbranched pathways. Figure 7.10 summarizes the inhibitory effects of the amino acids derived from aspartate. The activity of aspartokinase I and of homoserine dehydrogenase I is controlled by L-threonine. Aspartokinase III activity is regulated by L-lysine. Interestingly, the isoenzyme II is not subject to feedback inhibition by L-methionine; the latter inhibits the first enzyme specifically involved in methionine synthesis (O-succinyl homoserine synthetase).

Figure 7.11 summarizes the control of enzyme activity in the aromatic

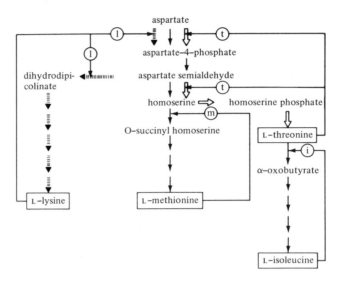

Figure 7.10. Feedback inhibition in the pathway leading from aspartate to the aspartate family of amino acids in *E. coli*. L-isoleucine inhibits threonine deaminase; L-threonine inhibits aspartokinase I and homoserine dehydrogenase I; L-methionine inhibits O-succinyl homoserine synthetase; L-lysine inhibits aspartokinase III and dihydrodipicolinate synthetase.

amino acid pathway of *E. coli*. One DAHP synthetase is inhibited by L-phenylalanine and the second one by L-tyrosine. The third enzyme which exhibits the lowest activity of the three enzymes, is not inhibited by tryptophan. The latter, however, inhibits anthranilate synthase. The two chorismate mutases are also under the control of the corresponding end-products. Most enterobacteria contain three DAHP synthetases, and these enzymes are regulated in a similar fashion. In *Bacillus* species, there is only one synthetase present; it is not inhibited by the end products of the pathway but by the branch point intermediates chorismate or prephenate. This type of control of activity is called **sequential feedback inhibition**. Still another pattern of regulation of the aromatic amino acid pathway was found in hydrogen-oxidizing bacteria such as *Alcaligenes eutrophus*. Here the DAHP synthetase is subject to **cumulative inhibition** by phenylalanine and tyrosine. Table 7.3 gives the percent inhibition observed with phenylalanine, tyrosine, and with the mixture of these amino acids. The inhibition rates are not additive. The residual activity observed in the presence of the mixture is equal to the product of the residual activities observed with phenylalanine and tyrosine. This is characteristic for cumulative inhibition.

From the examples of feedback inhibition given it is evident that in most cases the inhibitor differs significantly in its chemical structure from the substrates of the enzyme it acts upon. Carbamyl phosphate and aspartate, the substrates of aspartate transcarbamylase, are very different from the

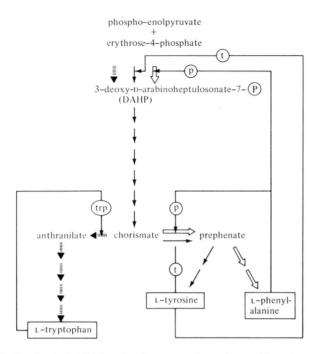

Figure 7.11. Feedback inhibition in the aromatic amino acid pathway of *E. coli*. L-phenylalanine inhibits DAHP synthetase I and chorismate mutase I; L-tyrosine inhibits DAHP synthetase II and chorismate mutase II; L-tryptophan inhibits anthranilate synthase.

inhibitor CTP. So it is very unlikely that CTP competes with the substrates of aspartate transcarbamylase for the active site of the enzyme as does malonate with succinate at succinate dehydrogenase. Since the inhibitors are not steric analogues of the substrates, enzymes must contain special binding sites for

Table 7.3. Regulation of DAHP synthetase activity of *Alcaligenes eutrophus* by cumulative inhibition[a]

amino acid present in the DAHP synthetase assay mixture	inhibition (%)	residual activity (100%=1) observed	calculated
phenylalanine	25	0.75	0.75
			×
tyrosine	47.5	0.525	$\overline{0.525}$
mixture	58.2	0.418	$\overline{0.394}$

[a]R. A. Jensen, D. S. Nasser, and E. W. Nester, *J. Bacteriol.* **94**, 1582–1593 (1967).

them. Monod, Changeux, and Jacob coined the term **allosteric sites** for these areas on the enzymes. Compounds that are bound at these sites and that alter the activity of the enzymes are called **allosteric effectors** and the enzymes subject to control by these effectors are called **allosteric enzymes**.

B.   Properties of allosteric enzymes

Allosteric enzymes are oligomers; they consist of 2, 4, 6, or more subunits which may be identical or not. One of the most frequent characteristics of allosteric enzymes is their anomalous kinetic behavior. If the reaction rate is plotted against the substrate concentration, a sigmoid curve is obtained and not a hyperbola as with enzymes that follow Michaelis–Menten kinetics. This is shown for aspartate transcarbamylase in Figure 7.12. CTP is a negative effector of this enzyme and it can be seen that at intermediate concentrations of aspartate the decrease in velocity caused by 0.2 mM CTP is considerable.

The sigmoidal response of the reaction velocity to an increase of the substrate concentration is explained by **cooperative binding**. In the absence of substrate, the active sites of enzymes are assumed to exist in forms which have little catalytic activity. However, through interaction with the substrate the active sites are assumed to be forced into a conformational state which is catalytically active and brings about the enzymatic reaction (induced-fit model of Koshland).

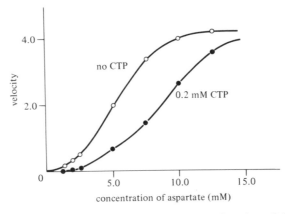

Figure 7.12. Activity of aspartate transcarbamylase as a function of the concentration of aspartate. [J. C. Gerhart and A. B. Pardee, *J. Biol. Chem.* **237**, 891–896 (1962).]

This model implies that the affinity of a particular active site for its substrate may also be subject to alteration. It could exist in two forms which have a different tendency to undergo the conformational change of the induced fit in the presence of substrate, and thus represent a low-affinity state or a high-affinity state of the enzyme. The increase of enzyme activity by cooperative binding would then mean that a substrate molecule bound to one subunit of an oligomeric enzyme forces the active sites of the other subunits from a low-affinity into a high-affinity state. Accordingly, a negative effector (CTP in Figure 7.12) bound at the allosteric site of one subunit not only keeps the active site of this subunit in the low-affinity state but also prevents the transition of the sites of the other subunits of the enzyme molecule from the low-affinity to the high-affinity state. A positive effector has the opposite effect.

Two models for cooperative binding have been proposed: the **sequential interaction model** of Koshland and co-workers and the **concerted symmetry model** of Monod and co-workers. The latter model assumes that the oligomeric enzyme exists as an equilibrium mixture of two forms: the T state in which all subunits are in the low-affinity form and the R state in which all subunits are in the high-affinity form.

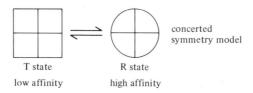

Effectors then change the equilibrium between these states. The sequential model of Koshland allows also intermediate states of the enzyme in which, for instance, two subunits of a tetrameric enzyme have a high affinity for the substrate and two have a low affinity.

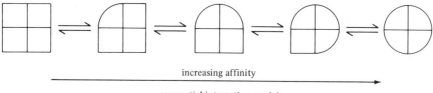

A positive effector would then increase the number of subunits in the high-affinity state and a negative effector would decrease this number.

C. Allosteric control of central pathways

The main objectives of catabolic and central pathways are to provide the cell with energy and with starting material for biosynthesis, and it is, therefore, reasonable that the regulatory signals used here for control are ultimate products of the energy metabolism and central precursors of the biosynthetic metabolism. Table 7.4 summarizes some allosteric enzymes involved in central pathways of *E. coli* and their inhibitors and activators. An increase of the $NADH_2$ concentration in the cells signals that the respiratory chain is saturated with $NADH_2$ and that the tricarboxylic acid cycle may slow down. Consequently, citrate synthase and also malate dehydrogenase and the pyruvate dehydrogenase complex are subject to inhibition by $NADH_2$. Moreover, citrate synthase is inhibited by α-oxoglutarate and the pyruvate dehydrogenase complex by acetyl-CoA.

Not all bacterial citrate synthases are inhibited by $NADH_2$. The $NADH_2$-inhibitable type of enzyme is found mainly in Gram-negative bacteria; Gram-positive bacteria contain a synthase that is inhibited by ATP like the enzyme of eukaryotic organisms. α-Oxoglutarate functions as inhibitor in enterobacteria only.

PEP carboxylase is inhibited by aspartate and malate. A high level of the latter compounds signals that C_4-dicarboxylic acids need not to be synthesized. An increase of the acetyl-CoA concentration, on the other hand, might indicate a shortage of C_4-dicarboxylic acids. Thus acetyl-CoA is an activator of PEP carboxylase. Many organisms contain pyruvate carboxylase instead of PEP carboxylase as anaplerotic enzyme. The pyruvate carboxylase from most sources is also activated by acetyl-CoA.

Table 7.4. Allosteric enzymes involved in central pathways of *E. coli*[a]

enzyme	inhibitor	activator
ADP-glucose pyrophosphorylase	AMP	glyceraldehyde-3-P, F-P$_2$,[b] PEP
fructose bisphosphatase	AMP	—
phosphofructokinase	PEP	ADP, GDP
pyruvate kinase		F-P$_2$
pyruvate dehydrogenase	NADH$_2$, acetyl-CoA	PEP, AMP, GDP
PEP carboxylase	aspartate, malate	acetyl-CoA, F-P$_2$, GTP, CDP
citrate synthase	NADH$_2$, α-oxoglutarate	
malate dehydrogenase	NADH$_2$	

[a] B. D. Sanwal, *Bacteriol. Rev.* **34**, 20–39 (1970).
[b] F-P$_2$, fructose-1,6-bisphosphate.

Fructose-1,6-bisphosphate is a strategic branch point of glycolysis and of glycogen formation. An increased AMP level inhibits ADP-glucose pyrophosphorylase and fructose bisphosphatase; both enzymes are involved in glycogen formation. If the concentration of fructose-1,6-bisphosphate ($F\text{-}P_2$) increases, this has a positive effect on glycolysis; both pyruvate kinase and PEP carboxylase are activated by $F\text{-}P_2$. Glycogen formation is also stimulated by activation of ADP-glucose pyrophosphorylase. If a further increase of the glycolytic activity is not desirable, PEP inhibits phosphofructokinase and glycogen formation is favored.

Figure 7.13 summarizes the control of glycolytic and glycogenic reactions in *E. coli*. It can be seen that the target enzymes for the control of reaction

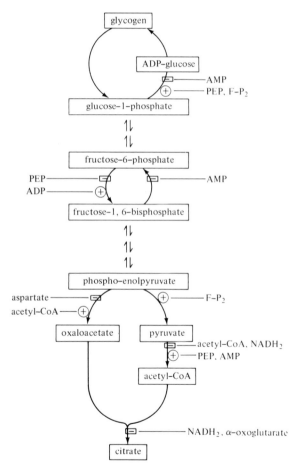

Figure 7.13. Schematic representation of glycolytic and glycogenic reactions in *E. coli* and their control. $F\text{-}P_2$, fructose-1,6-bisphosphate; ⊟—, inhibitor; ⊕—, activator.

sequences usually catalyze reactions that are irreversible under physiological conditions. These enzymes (phosphofructokinase, pyruvate kinase, etc.) function as pacemakers, and regulation is very efficient at these points. Furthermore, it is apparent from Figure 7.13 that antagonistic enzymes occur in the cells simultaneously, e.g., phosphofructokinase and fructose-1,6-bisphosphatase. Particularly these enzymes have to be under stringent control, otherwise **futile cycles** would be established, which bring about the hydrolysis of ATP as net reaction: fructose-6-phosphate is phosphorylated by phosphofructokinase, and the bisphosphate is hydrolyzed again by the phosphatase. A futile cycle could also be established with glucose-1-phosphate, ADP-glucose, and glycogen as participating metabolites.

Figure 7.13 shows also that adenine nucleotides are important effectors. AMP is formed in many biosynthetic reactions from ATP and so is ADP. Any increase of the concentration of these nucleotides leads to a stimulation of ATP-yielding reactions. By regulation of ATP-producing and-consuming reactions organisms try to maintain a constant energy status. According to Atkinson this energy status can be described by the **energy charge**, which is defined as:

$$ec = \frac{[ATP] + \frac{1}{2}[ADP]}{[ATP] + [ADP] + [AMP]}$$

Systems containing only ATP have an energy charge of 1 and for those containing only AMP the ec is zero. Measurements have shown that the energy charge of growing organisms is about 0.8. *E. coli* cells die at ec values below 0.5.

D. Covalent modification of enzymes

In recent years the importance of another mechanism for the regulation of enzyme activity has been recognized more and more: modulation of enzyme activity by covalent modification of enzymes. Whereas in allosteric regulation a low-molecular-weight compound (metabolite) is bound to or released from the enzyme, in the type of regulation discussed now the enzyme is covalently modified in an enzyme-catalyzed reaction. The principle of this regulatory mechanism is evident from the following equations:

$$\text{enzyme} - X \xrightarrow[\text{modifying enzyme(s)}]{} \text{enzyme} + X$$
$$\text{(active)} \qquad\qquad\qquad \text{(inactive)}$$

or

$$\text{enzyme} + X \xrightarrow[\text{modifying enzyme(s)}]{} \text{enzyme} - X$$
$$\text{(active)} \qquad\qquad\qquad \text{(inactive)}$$

The enzymes in question exist in two forms, an active one and an inactive one, and these forms are interconvertible.

Both forms differ in that one form is covalently substituted and the other is not. The first enzyme found by Cori and co-workers to exist in two forms was muscle **glycogen phosphorylase**. Form b is virtually inactive; it is present in resting aerobic muscle. When glycogenolysis becomes necessary in working muscle, form b of phosphorylase is phosphorylated and converted thereby into the highly active form a. At the expense of ATP, one phosphoryl group is linked to a serine residue per subunit. Conversion of form a into form b again is achieved by the hydrolytic removal of the phosphate groups. Thus for the modification of phosphorylase two enzymes are required: phosphorylase kinase and phosphorylase phosphatase. These regulatory enzymes, of course, also have to be under control, so that a highly sensitive cascade-like mechanism is responsible for the regulation of muscle phosphorylase.

Here it is necessary to mention that the glycogen phosphorylase of prokaryotes is not regulated by covalent modification and that enzyme systems modified by phosphorylation/dephosphorylation are generally more common among eukaryotes. Two interesting systems of covalent modification have been observed in bacteria: **glutamine synthetase** of *E. coli* is regulated by adenylylation/deadenylylation and **citrate lyase** of *Rhodopseudomonas gelatinosa* is regulated by acetylation/deacetylation. Figure 7.14 summarizes the control of glutamine synthetase as unraveled by Stadtman, Holzer, and

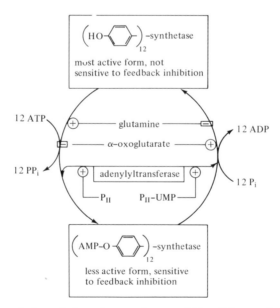

Figure 7.14. Regulation of glutamine synthetase of *E. coli* by covalent modification. The interconversion is catalyzed by an adenylyltransferase plus a regulatory protein P_{II}. Deadenylylation requires UTP for uridylylation of P_{II} to form P_{II}-UMP, α-oxoglutarate, and the absence of glutamine; adenylylation requires glutamine, low levels of α-oxoglutarate, and P_{II}.

their co-workers. This enzyme consists of 12 subunits. The deadenylylated form is very active; at high levels of glutamine and low levels of α-oxoglutarate the enzyme is adenylylated at one tyrosine residue per subunit. The resulting enzyme form exhibits low activity and, in addition, is subject to cumulative feedback inhibition by AMP, CMP, tryptophan, histidine, glucosamine-6-phosphate, and carbamyl phosphate. In case of a shortage of glutamine the AMP residues are transferred to inorganic phosphate and the active enzyme is restored again. The interconversion of the two forms of glutamine synthetase is catalyzed by an adenylyltransterase and a regulatory protein (P_{II}) which can be uridylylated by UTP. In the presence of the uridylylated protein (P_{11}-UMP), the transferase catalyzes the deadenylylation of the synthetase; otherwise, adenylylation is favored.

The glutamine synthetase from many other bacterial sources is not subject to regulation by covalent modification. Similarly, only the citrate lyase of the phototrophic bacterium *Rhodopseudomonas gela.inosa* has been found so far to be regulated by acetylation/deacetylation (Figure 7.15). Citrate lyase cleaves citrate to oxaloacetate and acetate. Active enzyme is present if citrate is available to the cells as substrate. Under these conditions the deacetylating enzyme is inhibited by L-glutamate for which citrate serves as precursor. In the absence of citrate, the glutamate concentration in the cells decreases and citrate lyase is inactivated by a specific deacetylase. If citrate becomes available again to the cells, the active enzyme is formed by acetylation. The citrate lyase system differs from the other systems discussed in that the substituent (the acetyl group) participates in the enzymatic reaction. The mechanism of the lyase reaction according to Eggerer and co-workers is as follows:

$$\text{acetyl-S-enzyme} + \text{citrate} \rightleftharpoons \text{citryl-S-enzyme} + \text{acetate}$$

$$\text{citryl-S-enzyme} \rightleftharpoons \text{acetyl-S-enzyme} + \text{oxaloacetate}$$

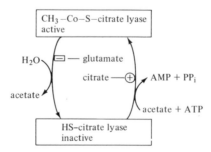

Figure 7.15. Regulation of citrate lyase of *Rhodopseudomonas gelatinosa* by acetylation and deacetylation. Deacetylation is inhibited by L-glutamate; acetylation proceeds in the presence of citrate.

The acetyl group participates in the initial transferase reaction and it is thus apparent that the HS-enzyme must be inactive.

III.   Summary

1. The process of substrate-mediated enzyme synthesis is termed enzyme induction. In coordinated induction all enzymes of a pathway are under the control of one inducer. In sequential induction several inducers are involved.

2. The genes of inducible enzymes which are coordinately formed are located on the bacterial genome adjacent to one another. Together with an operator and a promotor they form a regulatory unit, the operon. In the absence of the inducer gene expression is prevented by a repressor, which is bound at the operator. The inducer prevents that binding and thus allows transcription of the operon.

3. The phenomenon that readily utilizable substrates prevent the induction of the enzymes of other catabolic pathways is called catabolite repression. Transport of these substrates into the cells causes a drastic decrease of the intracellular level of cyclic AMP. The latter, however, is required for the initiation of RNA synthesis at operons that code for inducible pathways.

4. The level of the enzymes of anabolic pathways is regulated by end product repression. An end product, which accumulates in the cell, combines with a specific aporepressor to yield an active repressor. The latter prevents transcription of the genes involved in the synthesis of the enzymes of this particular pathway. Regulation of the enzyme levels in branched pathways is difficult; isoenzymes and multivalent repression help to avoid that one end product interferes with the synthesis of another end product of the same pathway.

5. In feedback inhibition, the end product of an anabolic pathway decreases the activity of the first enzyme specifically involved in its formation. This allows the rapid adjustment of the rate of product synthesis to the cellular demands. The target enzymes for feedback inhibition are allosteric enzymes. Besides their active sites, allosteric enzymes contain specific binding sites for their effectors (inhibitors and activators). Most allosteric enzymes show a sigmoidal response of the reaction velocity to an increase of the substrate concentration. This is explained by cooperative binding of the substrate and/ or the effectors.

6. Regulatory signals used for the control of the activity of central pathways are products of energy metabolism (AMP, ADP, ATP, $NADH_2$) and important precursors of biosynthetic metabolism (PEP, acetyl-CoA, aspartate). The enzymes under allosteric control usually function as pacemakers in central pathways.

7. The energy status of cells is described by the energy charge: $ec = [ATP] + \frac{1}{2}[ADP]/[ATP] + [ADP] + [AMP]$.

8. Some key enzymes are regulated by covalent modification. These enzymes exist in an active form and an inactive or less active form. Interconversion of these forms is achieved by phosphorylation/dephosphorylation (eukaryotic enzymes involved in glycogen metabolism), adenylylation/ deadenylylation (glutamine synthesis of *E. coli*), and acetylation/deacetylation (citrate lyase of *Rhodopseudomonas gelatinosa*).

Chapter 8

Bacterial Fermentations

Large areas in the soil, in rivers, lakes, and oceans are devoid of oxygen. Nevertheless, numerous microorganisms live in these anaerobic environments. We have already discussed the dissimilatory reduction of nitrate, which takes place under anaerobic conditions. Besides this process, two other microbiological processes account to a large extent for the biological activities in environments devoid of oxygen: bacterial fermentations and bacterial photosynthesis. The term fermentation was first defined by Pasteur, to whom we owe the pioneering work in this field; he described fermentations as life in the absence of oxygen. Today fermentations can be defined as those biological processes that occur in the dark and that do not involve respiratory chains with oxygen or nitrate as electron acceptors. In many fermentations ATP is formed only by substrate-level phosphorylation. However, in a number of fermentations electron transport phosphorylation is also involved in ATP synthesis.

The bacteria carrying out fermentations are either facultative or obligate anaerobes. Facultative anaerobes such as the enterobacteria grow as aerobic heterotrophs in the presence of oxygen; under anaerobic conditions they carry out a fermentative metabolism. In contrast, **obligate anaerobes** are not able to synthesize the components of an oxygen-linked respiratory chain. Consequently, they cannot grow as aerobes. Moreover, many of the obligate anaerobes do not even tolerate oxygen and are killed in air. These organisms are referred to as **strict anaerobes**.

When reduced flavoproteins or reduced iron-sulfur proteins come together with oxygen, two toxic compounds are formed: hydrogen superoxide and the superoxide radical (see Chapter 2, Section V). Aerobes contain catalase and superoxide dismutase for destruction of these compounds. Table 8.1 shows the activity of these enzymes as found in aerobes, aerotolerant an-

Table 8.1. Activity of superoxide dismutase and catalase in a variety of microorganisms[a]

organism	specific activity (U/mg)	
	superoxide dismutase	catalase
Aerobes		
Escherichia coli	1.8	6.1
Salmonella typhimurium	1.4	2.4
Rhizobium japonicum	2.6	0.7
Micrococcus radiodurans	7.0	289
Pseudomonas species	2.0	22.5
Aerotolerant anaerobes		
Eubacterium limosum	1.6	0
Streptococcus faecalis	0.8	0
Streptococcus lactis	1.4	0
Clostridium oroticum	0.6	0
Lactobacillus plantarum	0	0
Strict anaerobes		
Veillonella alcalescens	0	0
Clostridium pasteurianum	0	0
Clostridium barkeri	0	0
Clostridium sticklandii	0	0
Butyrivibrio fibrisolvens	0	0.1

[a]J. M. McCord, B. B. Keele, and I. Fridovich, *Proc. Natl. Acad. Sci.* **68**, 1024–1027 (1971).

aerobes, and strict anaerobes. It is evident that aerobes contain high levels of catalase and superoxide dismutase. Most aerotolerant anaerobes are devoid of catalase but contain superoxide dismutase. Strict anaerobes lack both enzymes. Although other factors might be involved, it can be concluded that aerointolerant species die in the presence of oxygen because of the deleterious effects of the superoxide radical.

For growth of most strict anaerobes it is not sufficient to exclude molecular oxygen. A low redox potential is required and usually the growth media have to be supplemented with SH-compounds such as thioglycolate, cysteine, or sodium sulfide. These compounds establish reducing conditions. Methanogenic bacteria grow only in media with a redox potential lower than $E'_0 = -0.3$ V.

A comparison of aerobic heterotrophic metabolism with fermentative metabolism reveals one major difference: aerobic heterotrophs couple the oxidation of organic substrates to the reduction of oxygen or nitrate. This involves respiratory chains with high ATP yields. With the exception of some incomplete oxidizers, the substrate is converted into cell material, carbon dioxide, and water.

Fermentative anaerobes carry out a variety of oxidation-reduction reactions involving organic compounds, carbon dioxide, molecular hydrogen, and sulfur compounds. All these reactions have in common that they can yield only little ATP. Therefore, the amount of cells obtained per mol of substrate is much smaller than with aerobes and, in addition to cell material, large amounts of "fermentation end products" are formed.

There are two fermentative processes that at first appear to be quite similar to oxygen and nitrate-dependent respirations: the reduction of CO_2 to methane and of sulfate to sulfide. Together with the nitrate-dependent respiration these processes are frequently called **anaerobic respirations**. However, they bear little resemblance to the process of denitrification. The reduction of CO_2 and of sulfate is carried out by strict anaerobes whereas nitrate reduction is carried out by aerobes if oxygen is not available. Nitrate respirers contain a true respiratory chain; sulfate and CO_2 reducers do not. Furthermore, the energetics of these processes are very different. Table 8.2 shows that the free energy change of O_2 and nitrate reduction per two electrons is about the same; the values are much lower for CO_2 and sulfate reduction. In fact, the values are so low that the formation of one ATP per H_2 or $NADH_2$ oxidized cannot be expected (ATP synthesis from $ADP + P_i$ requires more than 8.5 kcal(35.6kJ)/mol). Consequently, not all the reduction steps in methane and sulfide formation can be coupled to ATP synthesis. Only the reduction of one or two intermediates may yield ATP by electron transport phosphorylation, and the ATP gain is therefore small. In conclusion, it seems appropriate to reserve the term anaerobic respiration for denitrification and nitrate/nitrite respiration and to consider CO_2 reduction and sulfate reduction as fermentations.

Fermentations are usually classified according to the main fermentation end products, and we distinguish alcohol, lactate, propionate, butyrate, mixed acid, acetate, methane, and sulfide fermentations.

I. Alcohol Fermentation

A. Ethanol fermentation by yeasts

Alcohol fermentation is the domain of yeasts, notably of *Saccharomyces* species, and most of the ethanol formed in nature and produced by the fermentation industry comes from the anaerobic breakdown of glucose and other hexoses by these organisms. Gay-Lussac had already established in 1815 that hexoses are converted into ethanol and CO_2 according to the following equation:

$$C_6H_{12}O_6 \longrightarrow 2C_2H_5OH + 2CO_2$$

Figure 8.1 summarizes the alcohol fermentation as carried out by yeasts. It is apparent that yeasts employ the Embden–Meyerhof pathway for glucose

Table 8.2. Free energy changes in aerobic and anaerobic respiration and in sulfide and methane fermentation

redox reaction	$\Delta G'_0$ (kcal/mol acceptor)	$\Delta G'_0$ (kcal/$2e^-$)	
		H_2	$NADH_2$
$2H_2 + O_2 \rightarrow 2H_2O$	-113.38 (-474.38)	-56.69 (-237.19)	-52.36 (-219.07)
$\frac{5}{2}H_2 + NO_3^- + H^+ \rightarrow \frac{1}{2}N_2 + 3H_2O$	-133.99 (-560.61)	-53.60 (-224.24)	-49.27 (-206.12)
$4H_2 + SO_4^{2-} + \frac{3}{2}H^+ \rightarrow \frac{1}{2}H_2S + \frac{1}{2}HS^- + 4H_2O$	-36.68 (-153.46)	-9.17 (-38.36)	-4.84 (-20.24)
$4H_2 + CO_2 \rightarrow CH_4 + 2H_2O$	-31.26 (-130.79)	-7.81 (-32.70)	-3.48 (-14.58)

[a]kJ values in parentheses

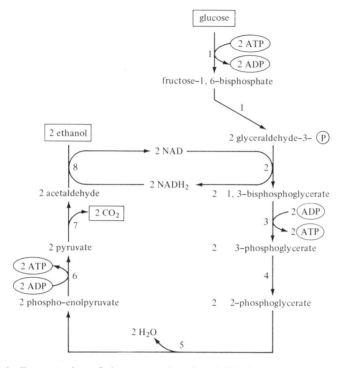

Figure 8.1. Fermentation of glucose to ethanol and CO_2 by yeasts. 1, enzymes of the Embden–Meyerhof pathway; 2, glyceraldehyde-3-phosphate dehydrogenase; 3, 3-phosphoglycerate kinase; 4, phosphoglycerate mutase; 5, enolase; 6, pyruvate kinase; 7, pyruvate decarboxylase; 8, alcohol dehydrogenase.

degradation. Thus, 2 mol of pyruvate are formed from 1 mol of glucose. Pyruvate, however, is not converted to acetyl-CoA as in aerobic metabolism, but is decarboxylated to acetaldehyde. The enzyme which catalyzes this reaction is **pyruvate decarboxylase**; it can be regarded as the key enzyme of alcohol fermentation. Pyruvate decarboxylase contains bound thiamin pyrophosphate; in fact, the function of thiamin pyrophosphate was first studied using this enzyme. As in the pyruvate dehydrogenase reaction, hydroxyethyl-thiamin pyrophosphate enzyme is an intermediate. The hydroxyethyl group, however, is not oxidized but is released as free acetaldehyde. The latter is then reduced to ethanol by alcohol dehydrogenase.

One important feature of fermentations is apparent from Figure 8.1: an even hydrogen balance. Since there is no external H-acceptor like oxygen, $NADH_2$-producing and $NADH_2$-consuming reactions have to be balanced out. This also implies that $NADH_2$ formation is avoided by anaerobes if possible. Therefore, they do not oxidize acetyl-CoA via the tricarboxylic acid cycle. The cycle is usually interrupted between α-oxoglutarate and succinyl-CoA so that glutamate can still be formed from oxaloacetate and acetyl-CoA via citrate.

$$\text{hydroxyethyl-thiamin pyrophosphate enzyme}$$

hydroxyethyl-thiamin pyrophosphate enzyme

B. The Pasteur effect

The net ATP yield of the alcohol fermentation is 2 mol ATP/mol glucose (Figure 8.1)—much lower than the ATP yield of aerobic metabolism. Yeast cells respond to this large difference. When they are transferred from aerobic to anaerobic conditions they increase the rate of glucose breakdown by a factor of 3 to 4; a change from anaerobic to aerobic metabolism is accompanied by a reduction of this rate and a stoppage of alcohol formation. This phenomenon is called the **Pasteur effect**. Apparently, the Pasteur effect is the result of differences in the cell's energy charge under aerobic and anaerobic conditions. In the presence of oxygen the respiratory chain and the sites of substrate-level phosphorylation in the glycolytic pathway compete for ADP. Furthermore, phosphofructokinase activity is controlled by the level of ATP and citrate. Under anaerobic conditions the activity of phosphofructokinase increases because it is activated by ADP and AMP; also, more ADP is available for the enzymes catalyzing substrate-level phosphorylation reactions. All this allows a greater substrate flow through the glycolytic pathway.

C. Alcohol fermentation by bacteria

It should be mentioned that yeasts are not truly facultatively anaerobic organisms. They grow only for some generations under these conditions. There are, however, some bacterial species that carry out an alcohol fermentation and grow very well under anaerobic conditions. *Zymomonas mobilis* isolated from Mexican pulque and the closely related *Zymomonas anaerobica* degrade glucose to pyruvate via the Entner–Doudoroff pathway; they contain pyruvate decarboxylase and form nearly 2 mol each of ethanol and carbon dioxide from 1 mol glucose. *Sarcina ventriculi*, a strict anaerobe capable of growth under extremely acidic conditions, and *Erwinia amylovora*, a facultatively anaerobic enterobacterium, ferment glucose to ethanol and CO_2 via the Embden–Meyerhof pathway and the pyruvate decarboxylase and alcohol dehydrogenase reactions. Both organisms, however, form small quantities of other products in addition: acetate and molecular hydrogen (*S. ventriculi*) and lactate (*E. amylovora*). In general, pyruvate decarboxylase is rare in bacteria. Many lactic acid bacteria, enterobacteria, and clostridia form considerable amounts of ethanol but do not employ pyruvate decarboxylase for acetaldehyde synthesis. In these organisms acetyl-CoA functions as ultimate precursor of acetaldehyde; it is reduced by acetaldehyde dehydrogenase:

$$CH_3-CO-CoA + NADH_2 \xrightarrow[\text{dehydrogenase}]{\text{acetaldehyde}} CH_3-C\underset{H}{\overset{O}{\backslash}} + CoA + NAD$$

The investigation of the alcohol fermentation has been of inestimable importance for the development of biological sciences. In 1892 the Buchner brothers discovered that an extract of macerated yeast fermented glucose to ethanol. This demonstrated that a complex biological process could function outside the cell.

In 1905 Harden and Young discovered the importance of phosphate esters in metabolism; they identified a hexose bisphosphate (fructose 1,6-bisphosphate) as an intermediate in sugar fermentation. Later, the path of glucose breakdown and the function of the coenzymes in this process were unraveled by Neuberg, Embden, Meyerhof, Parnas, and Warburg.

II. Lactate Fermentation

Lactate is a very common end product of bacterial fermentations. Some genera—often referred to as **lactic acid bacteria**—form large amounts of this product. These microorganisms have in common that they are highly saccharolytic and that they lack most anabolic pathways. So they exhibit very complex nutritional requirements, which are met by their environment such as plant materials, milk, and the intestinal tract of animals. Most lactic acid bacteria are strictly fermentative but are aerotolerant. Some streptococci, however, can use oxygen as H-acceptor and even form cytochromes under certain conditions.

In Table 8.3 species of the genera *Lactobacillus, Sporolactobacillus, Streptococcus, Leuconostoc, Pediococcus,* and *Bifidobacterium* are listed. Either of three pathways is employed by these microorganisms for the fermentation of carbohydrates to lactate.

The **homofermentative pathway** yields 2 mol of lactate per mol of glucose:

$$\text{glucose} \xrightarrow{\text{homofermentative pathway}} 2 \text{ lactate}$$

The **heterofermentative pathway** yields 1 mol each of lactate, ethanol, and CO_2 per mol of glucose:

$$\text{glucose} \xrightarrow{\text{heterofermentative pathway}} \text{lactate} + \text{ethanol} + CO_2$$

The **bifidum pathway** yields acetate and lactate in a ratio of 3 to 2:

$$2 \text{ glucose} \xrightarrow{\text{bifidum pathway}} 3 \text{ acetate} + 2 \text{ lactate}$$

These equations are, of course, not followed completely by the lactic acid bacteria but in homofermenting organisms the yield is frequently 1.8 mol lactate/mol glucose. The heterofermenting bacteria produce 0.8 mol lactate and some acetate in addition to ethanol and CO_2.

Table 8.3. Homo- and heterofermentative lactic acid bacteria and configuration of lactic acid produced

genera and species	homo-fermentative	hetero-fermentative	configuration of lactic acid
Lactobacillus			
L. delbrueckii	+	−	D (−)
L. lactis	+	−	D (−)
L. bulgaricus	+	−	D (−)
L. casei	+	−	L (+)
L. plantarum	+	+	DL
L. curvatus	+	−	DL
L. brevis	−	+	DL
L. fermentum	−	+	DL
Sporolactobacillus			
S. inulinus	+	−	D (−)
Streptococcus			
S. faecalis	+	−	D (−)
S. cremoris	+	−	D (−)
S. lactis	+	−	D (−)
Leuconostoc			
L. mesenteroides	−	+	D (−)
L. dextranicum	−	+	D (−)
Pediococcus			
P. cerevisiae	+	−	DL
Bifidobacterium			
B. bifidum	−	−	L (+)

A. Homofermentative pathway

The homofermentative pathway is illustrated in Figure 8.2. A close relationship to the alcohol fermentation is apparent. Glucose is degraded via the Embden–Meyerhof pathway to pyruvate. The latter is, however, not decarboxylated to acetaldehyde as in the alcohol fermentation but used

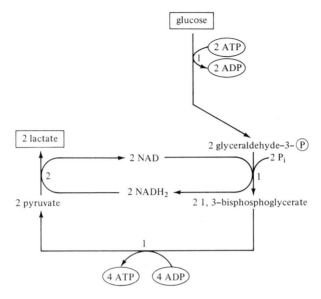

Figure 8.2. Formation of lactate from glucose by the homofermentative pathway.
1, enzymes of the Embden–Meyerhof pathway; 2, lactate dehydrogenase.

directly as H-acceptor. The ATP yield in both fermentations is the same,
2 ATP/glucose.

B. Heterofermentative pathway

The heterofermentative pathway is illustrated in Figure 8.3. As in the
oxidative pentose phosphate cycle, ribulose-5-phosphate is formed via 6-
phosphogluconate. Epimerization yields xylulose-5-phosphate, which is
cleaved into glyceraldehyde-3-phosphate and acetyl phosphate by an enzyme
not mentioned thus far, **phosphoketolase**. It contains thiamin pyrophosphate
and the formation of acetyl phosphate can be understood when enzyme-
bound α-hydroxyvinyl-thiamin pyrophosphate is assumed to occur as an
intermediate.

Acetyl phosphate is converted into acetyl-CoA by phosphotransacetylase.
Subsequent reduction by acetaldehyde and alcohol dehydrogenases yields
ethanol. The glyceraldehyde-3-phosphate formed in the phosphoketolase
reaction is converted to lactate as in the homofermentative pathway.

In the course of this fermentation $2 NADH_2$ are formed and consumed;
the ATP yield is one per mol of glucose—i.e half of that of the homofermentative pathway.

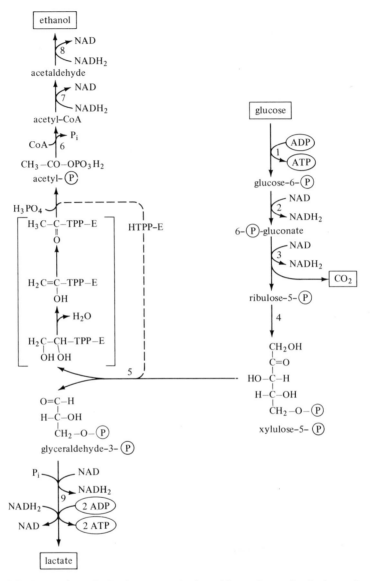

Figure 8.3. Formation of CO_2, lactate, and ethanol from glucose by the heterofermentative pathway. 1, hexokinase; 2, glucose-6-phosphate dehydrogenase; 3, 6-phosphogluconate dehydrogenase; 4, ribulose-5-phosphate 3-epimerase; 5, phosphoketolase. The cleavage reaction yields glyceraldehyde-3-phosphate and enzyme-bound α,β-dihydroxyethylthiamin pyrophosphate. This is converted to acetyl-TPP-E via the α-hydroxyvinyl derivative; phosphorylytic cleavage results in acetyl phosphate formation. 6, phosphotransacetylase; 7, acetaldehyde dehydrogenase; 8, alcohol dehydrogenase; 9, enzymes as in homofermentative pathway.

C. Bifidum pathway

In glucose breakdown by *Bifidobacterium bifidum* two phosphoketolases are involved: one specific for fructose-6-phosphate and one specific for xylulose-5-phosphate. The mechanism of both reactions is similar; fructose-6-phosphate phosphoketolase splits fructose-6-phosphate into acetyl phosphate and erythrose-4-phosphate.

$$
\begin{array}{ccc}
\text{CH}_2\text{OH} & & \text{CH}_3-\text{CO}-\text{OPO}_3\text{H}_2 \\
| & & \\
\text{C}=\text{O} & & + \\
| & & \\
\text{HO}-\text{C}-\text{H} \quad \text{P}_i & \nearrow & \text{H}-\text{C}=\text{O} \\
| & & | \\
\text{H}-\text{C}-\text{OH} & & \text{H}-\text{C}-\text{OH} \\
| & \searrow & | \\
\text{H}-\text{C}-\text{OH} & & \text{H}-\text{C}-\text{OH} \\
| & & | \\
\text{CH}_2-\text{O}-\text{P} & & \text{CH}_2-\text{O}-\text{P}
\end{array}
$$

The bifidum pathway is illustrated in Figure 8.4. It exhibits a very interesting sequence of reactions. Without the participation of hydrogenation and dehydrogenation reactions 2 mol of glucose are converted into 3 mol of acetate and 2 mol of glyceraldehyde-3-phosphate. The conversion of the latter to lactate involves then glyceraldehyde-3-phosphate and lactate dehydrogenases. The formation of acetate from acetyl phosphate is coupled to the formation of ATP from ADP.

$$
\text{CH}_3-\underset{\underset{\text{O}}{\|}}{\text{C}}-\text{O}-\text{PO}_3\text{H}_2 + \text{ADP} \xrightleftharpoons{\text{acetate kinase}} \text{CH}_3-\text{COOH} + \text{ATP}
$$

This reaction, which is catalyzed by acetate kinase, is of great importance for all anaerobes that form acetate because it effects ATP synthesis by substrate-level phosphorylation. With 2.5 mol of ATP per mol of glucose the ATP yield of the bifidum pathway is higher than that of the homo- and heterofermentative pathway.

D. Stereospecificity of lactate dehydrogenases

Table 8.3 has shown that lactic acid bacteria form either D(−)-, L(+)-, or DL-lactic acid. D(−)-Lactic acid-formers produce this enantiomer exclusively whereas L(+)-lactic acid-formers always produce some D(−)-enantiomer. Most organisms which excrete DL-lactate contain two lactate dehydrogenases which differ in their stereospecificity. Some lactobacilli, however, produce first L(+)-lactic acid, which—while accumulating—induces a racemase. This enzyme then converts the L(+)-form into the D(−)-form until equilibrium is reached. *L. curvatus* belongs to this rather small group of lactobacilli.

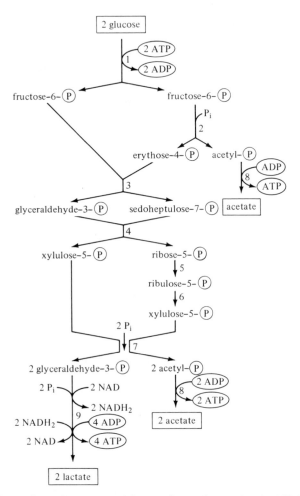

Figure 8.4. Formation of acetate and lactate from glucose by the bifidum pathway. 1, hexokinase and glucose-6-phosphate isomerase; 2, fructose-6-phosphate phospho-ketolase; 3, transaldolase; 4, transketolase; 5, ribose-5-phosphate isomerase; 6, ribulose-5-phosphate 3-epimerase; 7, xylulose-5-phosphate phosphoketolase; 8, acetate kinase; 9, enzymes as in homofermentative pathway.

The lactate dehydrogenases of streptococci, of *Bifidobacterium bifidum*, and of *L. casei* have an absolute and specific requirement for fructose 1,6-bisphosphate and manganese ions. The bisphosphate functions as positive allosteric effector. Lactate dehydrogenases from other lactic acid bacteria do not show such a requirement.

E. Fermentation of other saccharides

A large number of other sugars besides glucose are fermented by lactic acid bacteria; they include fructose, galactose, mannose, saccharose, lactose, maltose, and pentoses. Certain variations of the normal fermentation schemes have been observed with some of these substrates. Pentoses are, for instance, fermented by some homofermentative lactobacilli (e.g., *L. casei*) using phosphoketolase—the key enzyme of the heterofermentative pathway. Fructose is fermented by *Leuconostoc mesenteroides*, but part of the $NADH_2$ formed is not used to reduce acetyl-CoA to ethanol; instead fructose is reduced to mannitol. Thus, the products of fructose fermentation are lactate, acetate, CO_2, and mannitol.

F. Malo-lactate fermentation

The decrease of the acidity of wine is partly due to the conversion of L-malate to L-lactate. This process is called malo-lactate fermentation, and it is carried out by some lactic acid bacteria, e.g., *L. plantarum*, *L. casei*, and *Lc. mesenteroides*. In the presence of L-malate and a fermentable sugar these organisms synthesize a special **malic enzyme,** which converts L-malate into L-lactate:

$$\text{L-malate} \xrightarrow{\text{Mn}^{2+}, \text{ NAD}} \text{L-lactate} + CO_2$$

The purified enzyme does not possess lactate dehydrogenase activity so that free pyruvate cannot be an intermediate. Presumably enzyme-bound oxaloacetate and pyruvate function as intermediates in this reaction. The malo-lactate fermentation does not yield any ATP; consequently the organisms grow only if a fermentable sugar is available in addition to L-malate.

G. Formation of diacetyl and acetoin

In addition to the usual end products of the lactate fermentation *Streptococcus cremoris* and *Leuconostoc cremoris* form acetoin and diacetyl—the characteristic flavor of butter. Citrate, which occurs in milk in concentrations up to 1 g/liter, is the preferred substrate for diacetyl formation by the above organisms. As is illustrated in Figure 8.5 citrate is cleaved by citrate lyase into acetate and oxaloacetate. The enzyme **citrate lyase** is the key enzyme for the anaerobic breakdown of citrate. It occurs in lactic acid bacteria, in enterobacteria, *Veillonella* species, *Clostridium sphenoides*, and *Rhodopseudomonas gelatinosa* but is not involved in citrate degradation under aerobic conditions; its interesting reaction mechanism and its regulation in *R. gelatinosa* have been discussed in Chapter 7, Section II.

The acetate formed in the citrate lyase reaction is excreted, and oxaloacetate is decarboxylated to yield pyruvate. Diacetyl synthesis does not proceed via α-acetolactate as in bacilli (see Figure 6.24) or in enterobacteria (Figure 8.13), but is accomplished by reaction of acetyl-CoA with "active

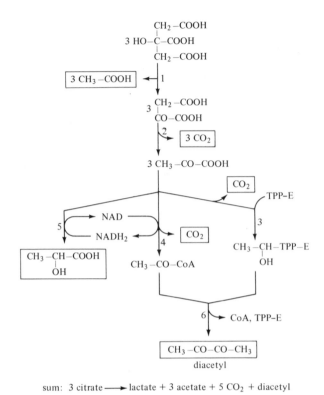

Figure 8.5. Formation of diacetyl from citrate by lactic acid bacteria. 1, citrate lyase; 2, oxaloacetate decarboxylase; 3, enzyme not characterized; 4, pyruvate dehydrogenase complex; 5, lactate dehydrogenase; 6, diacetyl synthase. Instead of pyruvate dehydrogenase the other pyruvate-degrading enzymes mentioned in the text may be involved as well.

acetaldehyde" (enzyme-bound hydroxyethyl-thiamin pyrophosphate). Reduction of diacetyl by acetoin dehydrogenase leads to acetoin:

$$CH_3-CO-CO-CH_3 + NADH_2 \xrightleftharpoons[\text{dehydrogenase}]{\text{acetoin}} CH_3-\underset{\underset{OH}{|}}{C}H-CO-CH_3 + NAD$$

Citrate is a good substrate for diacetyl formation because it yields pyruvate without the production of $NADH_2$, and pyruvate may thus be used for reactions other than lactate formation. Diacetyl synthesis requires the conversion of pyruvate into C_2-compounds. Moreover, all homofermentative lactic acid bacteria have to synthesize acetyl-CoA for biosynthetic purposes from pyruvate. The occurrence of three enzyme systems which yield acetyl-CoA or acetyl phosphate from pyruvate have been demonstrated in lactic acid bacteria. Streptococci contain the pyruvate dehydrogenase multienzyme complex. In addition, the pyruvate-formate lyase system, which will be discussed in Section IV of this chapter, has been detected in *S. faecalis*, in *Bifidobacterium bifidum*, and in some lactobacilli (e.g., *L. casei*). Finally, *L. delbrückii* and *L. plantarum* carry out a dismutation of pyruvate to lactate and acetyl phosphate:

$$\text{pyruvate} + P_i + FAD \xrightarrow{\text{pyruvate oxidase}} \text{acetyl phosphate} + CO_2 + FADH_2$$

$$FADH_2 + \text{pyruvate} \xrightarrow{\text{lactate oxidase}} \text{lactate} + FAD$$

Pyruvate oxidase is a flavoprotein that contains thiamin pyrophosphate. Neither lipoate nor coenzyme A are involved in acetyl phosphate formation. Lactate oxidase is also a flavoprotein, and the hydrogen transfer from one flavoprotein to the other is mediated by riboflavin.

H. Growth yields

It has been outlined in Chapter 3 (see Table 3.2) that approximately 35 mmol of ATP are necessary for the synthesis of 1 g of cells from glucose and inorganic salts. Thus, about 30 g of cells can be formed per mol of ATP. This value does not account for the energy of maintenance and for other ATP expenditures of the living cell. However, if the amount of ATP required for these processes is constant or if the variations are small, a given amount of ATP should result in the formation of the same amount of cells independent of the species. That this is true was shown by Bauchop and Elsden. They determined a yield of 22 g *S. faecalis* cells per mol of glucose fermented (all monomers required for macromolecule synthesis were supplied with the medium). Since the homofermentative pathway yields 2 ATP per glucose fermented, 11 g of cells were produced per mol of ATP:

$$\begin{aligned} S. \textit{ faecalis} \text{ growth yield}: Y_m &= 22 \text{ g/mol glucose} \\ Y_{ATP} &= 11 \text{ g/mol ATP} \end{aligned}$$

Zymomonas mobilis, which ferments glucose to ethanol and CO_2 via the Entner–Doudoroff pathway (1 ATP per glucose; see Chapter 8, Section I), yielded 8.3 g of cells per mol of glucose.

$$\begin{aligned} Z. \textit{ mobilis} \text{ growth yield}: Y_m &= 8.3 \text{ g/mol glucose} \\ Y_{ATP} &= 8.3 \text{ g/mol ATP} \end{aligned}$$

Thus, an organism gaining half the amount of ATP from glucose (compared to *S. faecalis*) formed approximately half the amount of cells. A Y_{ATP} value of 10.5 for growth with all the monomers available for biosynthesis is now generally accepted.

I. Growth in air

It has already been mentioned that many lactic acid bacteria are aerotolerant. These organisms contain superoxide dismutase but lack a true catalase. Instead, peroxidases are present, which catalyze the oxidation of organic compounds (alcohols, aldehydes) or of $NADH_2$ with H_2O_2:

$$NADH_2 + H_2O_2 \xrightarrow{\ NADH_2\ peroxidase\ } NAD + 2H_2O$$

However, some lactic acid bacteria, notably *S. faecalis*, *L. brevis*, and *L. plantarum* form a true catalase when they are grown in the presence of hemin. Thus, they are only unable to synthesize the heme prosthetic group of catalase.

When aerotolerant lactic acid bacteria are grown in air, the growth yield is usually much higher than under anaerobic conditions. For *S. faecalis* Y_m increases from 22 to 52. This increase is partly due to a change of the fermentation pattern: $NADH_2$ is oxidized with oxygen; pyruvate can be converted to acetyl-CoA and additional ATP can be formed by the acetate kinase reaction. Thus, oxygen changes the lactate fermentation into an acetate fermentation. There are, however, also indications that *S. faecalis* is capable of oxidative phosphorylation to a limited extent. In the presence of hemin and in air cytochromes are formed by this organism and by some other streptococci. So it appears that some lactic acid bacteria are not truly obligately anaerobic bacteria.

III. Butyrate and Butanol-Acetone Fermentation

The fermentation of sugars to butyric acid was discovered by Pasteur in 1861. Soon after, microorganisms responsible for the formation of butyrate were isolated, and it was found that several clostridial species carried out this type of fermentation. Generally, only obligate anaerobes form butyrate as a main fermentation product; they belong to the four genera *Clostridium*, *Butyrivibrio*, *Eubacterium*, and *Fusobacterium* (Table 8.4).

Table 8.4. Some species forming butyrate as a major fermentation end product

---

Clostridium butyricum
C. kluyveri
C. pasteurianum
Butyrivibrio fibrisolvens
Eubacterium limosum
   (formerly *Butyribacterium rettgeri*)
Fusobacterium nucleatum

The mechanism of butyrate formation was not well understood until Barker and collaborators did their pioneering studies on *C. kluyveri* and until ferredoxin was discovered in *C. pasteurianum*. The clostridia employ phosphotransferase systems for sugar uptake and the Embden–Meyerhof pathway for degradation of hexose phosphates to pyruvate. The conversion of pyruvate to acetyl-CoA involves an enzyme system not discussed thus far: pyruvate-ferredoxin oxidoreductase.

A. Ferredoxin and pyruvate-ferredoxin oxidoreductase

When a cell extract of *C. pasteurianum* is incubated with pyruvate under anaerobic conditions the decomposition of pyruvate according to the following equation can be observed:

$$CH_3-CO-COOH + H_3PO_4 \longrightarrow CH_3-CO-OPO_3H_2 + CO_2 + H_2$$

Acetyl phosphate is formed and hydrogen gas is evolved. This reaction is known as the **phosphoroclastic reaction**. A thorough investigation of the enzymes involved resulted in the finding that this phosphorylytic cleavage is the sum of several reactions; they are summarized in Figure 8.6. Pyruvate is first decarboxylated by pyruvate-ferredoxin oxidoreductase; the remaining C_2-moiety is covalently bound to the TPP-containing enzyme as in the pyruvate decarboxylase and the pyruvate dehydrogenase reactions. In the next step acetyl-CoA is formed; in contrast to pyruvate dehydrogenase the two hydrogens are not transferred to NAD, but are used to reduce ferredoxin. This difference is important because ferredoxin has a very low redox potential; at pH 7.0 it is about the same as that of the hydrogen electrode:

$$\text{hydrogen electrode} \quad E_0' = -0.42 \text{ V}$$
$$\text{ferredoxin} \quad E_0' = -0.41 \text{ V}$$
$$\text{NADH}_2 \quad E_0' = -0.32 \text{ V}$$

1    $\boxed{\text{pyruvate}}$ + TPP–E $\rightleftharpoons$ HETPP + $\boxed{CO_2}$

2 HETPP–E + ferredoxin + CoA $\rightleftharpoons$ TPP–E + acetyl–CoA + ferredoxin · H_2

hydrogenase
3   ferredoxin · H_2 $\rightleftharpoons$ ferredoxin + $\boxed{H_2}$

phosphotransacetylase
4   acetyl–CoA + P_i $\rightleftharpoons$ $\boxed{\text{acetyl phosphate}}$ + CoA

Figure 8.6. Steps of phosphoroclastic reaction. Steps 1 and 2 involve the enzyme pyruvate-ferredoxin oxidoreductase and ferredoxin. TPP-E, thiamin pyrophosphate-containing oxidoreductase; HETPP-E, hydroxyethyl-TPP-E. Steps 3 and 4 are catalyzed by hydrogenase and phosphotransacetylase, respectively.

Consequently, even in an environment saturated with hydrogen gas, reduced ferredoxin can transfer electrons to hydrogenase and hydrogen can be evolved. This is what is observed when clostridia ferment carbohydrates. Moreover, the pyruvate-ferredoxin oxidoreductase is reversible and under appropriate conditions pyruvate can be synthesized from acetyl-CoA, reduced ferredoxin, and CO_2 (see *C. kluyveri*).

$NADH_2$ is a much weaker reducing agent than reduced ferredoxin. This is one reason why the pyruvate dehydrogenase reaction is not reversible.

Ferredoxin was discovered in *C. pasteurianum* by Mortenson, Valentine, and Carnahan in 1962. It belongs to the **iron-sulfur proteins** (until recently called nonheme-iron proteins) and it should be remembered that this class of proteins plays an important role in the respiratory chain (see Figures 2.6 and 2.10). Clostridial ferredoxins have a molecular weight of 6,000 and contain eight iron atoms per molecule, which are arranged in the form of two cubanelike iron-sulfur clusters. One of these clusters is shown in Figure 8.7. The iron atoms are bound to cysteine residues of the peptide chain and they are interconnected by sulfide bridges. So ferredoxin contains eight sulfide groups; upon acidification of ferredoxin solutions they are liberated as H_2S.

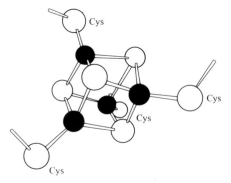

Figure 8.7. Iron-sulfur cluster of ferredoxin. ●, Fe atoms; ○, sulfur atoms.

This sulfur is therefore called labile sulfur. Clostridial ferredoxins are two-electron carriers (one electron per cluster). Because of their brownish color they are easily recognized in cell extracts and during fractionation of such extracts.

It should be emphasized that not all ferredoxins have the same general composition. The span of variation is apparent from Table 8.5. The ferredoxins of most clostridia and phototrophic bacteria (*Chromatium*) contain two iron-sulfur clusters. The ferredoxin of sulfate-reducing bacteria contains only one cluster per molecule and plants contain one cluster with two irons and two sulfide bridges. All ferredoxins have in common that they are low potential electron carriers.

One other protein is known that shares the latter property but differs in other aspects: **flavodoxin**. It is formed by several obligate anaerobic bacteria when they grow in iron-limited media. Iron and labile sulfide are absent and

Table 8.5. Properties of different ferredoxins

Organism	molecular weight	iron/ molecule	labile S/ molecule	$E_0'(V)$
Clostridium pasteurianum	6,000	8	8	−0.41
Chromatium strain D	10,000	8	8	−0.49
Desulfovibrio desulfuricans	6,000	4	4	−0.33
Spinach	11,000	2	2	−0.42

flavin mononucleotide functions as redox group. The redox potential lies around −0.4 V. Finally it should be mentioned that **rubredoxin** occurs in many anaerobes. This redox carrier contains per redox center one iron, which is bound to four cysteine residues of the polypeptide chain. Thus, rubredoxin does not contain labile sulfide. Its redox potential lies around −0.06 V. The function of rubredoxin in anaerobes is unknown; in some aerobes it is involved in alkane oxidation (see Figure 6.11).

B. The path of butyrate formation

The reactions involved in butyrate formation from glucose are summarized in Figure 8.8. The Embden–Meyerhof pathway and pyruvate-ferredoxin oxidoreductase yield two acetyl-CoA, two CO_2, and two reduced ferredoxins, which are reconverted to the oxidized form by hydrogenase under hydrogen evolution. Furthermore, two NAD are reduced and two ATP are formed. The advantage of hydrogen evolution in the course of pyruvate breakdown is apparent; only two $NADH_2$ are formed in the degradation of glucose to acetyl-CoA.

The $NADH_2$ formed in the glyceraldehyde-3-phosphate dehydrogenase reaction is oxidized by the conversion of two acetyl-CoA into butyrate. In this pathway acetoacetyl-CoA, L(+)-β-hydroxybutyryl-CoA, and crotonyl-CoA are intermediates. It should be remembered that the storage material poly-β-hydroxybutyric acid is a polymer of the D(−)-form (see Figure 5.15) and that the thioester of the D(−)enantiomer is an intermediate in long-chain fatty acid synthesis.

Butyrate is not formed from butyryl-CoA by simple hydrolysis; this would be a waste of energy. Instead the thioester linkage is transferred to acetate, where it can be used for ATP synthesis or anabolic purposes.

$$\text{butyryl-CoA} + \text{acetate} \xrightleftharpoons{\text{CoA transferase}} \text{acetyl-CoA} + \text{butyrate}$$

CoA transferase reactions are of general importance in fermentation processes.

Figure 8.8 shows that the ATP yield of the butyrate fermentation is 3 ATP/glucose; this is more than was gained in the fermentations discussed thus far.

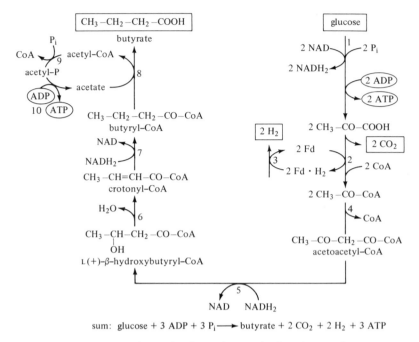

sum: glucose + 3 ADP + 3 P_i ⟶ butyrate + 2 CO_2 + 2 H_2 + 3 ATP

Figure 8.8. Path of butyrate formation from glucose. 1, phosphotransferase system and Embden–Meyerhof pathway; 2, pyruvate-ferredoxin oxidoreductase; 3, hydrogenase; 4, acetyl-CoA-acetyltransferase (thiolase); 5, L(+)-β-hydroxybutyryl-CoA dehydrogenase; 6, L-3-hydroxyacyl-CoA hydrolyase (crotonase); 7, butyryl-CoA dehydrogenase; 8, CoA-transferase; 9, phosphotransacetylase; 10, acetate kinase.

C. The formation of acetone and butanol

Figure 8.9 illustrates the course of product formation during growth of *C. acetobutylicum*. Toward the middle of the fermentation when the pH of the medium drops to 4.5 two new products are formed: acetone and butanol. Their formation coincides with a decrease in the concentration of butyrate. Apparently the fermentation is shifted toward neutral compounds in order to prevent a further decrease of pH. This is also indicated by the observation that solvent formation is largely reduced when the pH of the medium is kept above 5.

The reactions involved in acetone and butanol formation are summarized in Figure 8.10. At low pH, enzymes become active that convert acetoacetyl-CoA partly to acetone. Therefore, less acetoacetyl-CoA for NAD regeneration is available, and butyryl-CoA has to be reduced further to butanol. Additional butanol is formed from butyrate, which is taken up again from the medium. It is converted into butyryl-CoA by a CoA transferase reaction. H_2 initially evolved is used to provide $NADH_2$ for reduction to butanol.

Besides *C. acetobutylicum* a number of other clostridia also produce butanol and acetone. Some of them, including *C. butylicum*, reduce acetone to isopropanol.

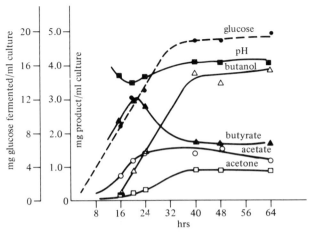

Figure 8.9. Product formation from glucose by *C. acetobutylicum*. [R. Davies and M. Stephenson, *Biochem. J.* **35**, 1320–1331 (1941).]

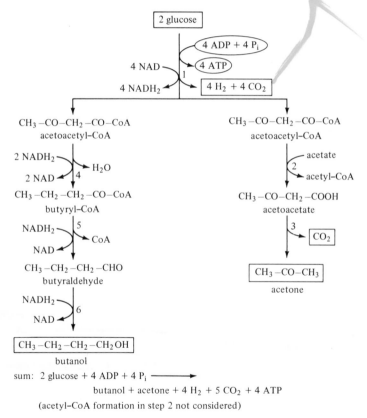

Figure 8.10. Formation of acetone and butanol by *C. acetobutylicum*. 1, reactions as in Figure 8.8; 2, CoA transferase (not definitely shown to be present in *C. acetobutylicum* but present in other clostridia); 3, acetoacetate decarboxylase; 4, L(+)-β-hydroxy-butyryl-CoA dehydrogenase, crotonase, and butyryl-CoA dehydrogenase; 5, butyraldehyde dehydrogenase; 6, butanol dehydrogenase.

D. Fermentation balances

Table 8.6 gives fermentation balances for *C. butyricum*, *C. perfringens* and *C. acetobutylicum*. It can be seen that the fermentation schemes depicted in Figures 8.8 and 8.10 are not exactly followed. *C. butyricum* forms acetate and some extra hydrogen; *C. perfringens* forms lactate and ethanol as do many saccharolytic clostridia. In the *C. acetobutylicum* fermentation the amount of H_2 evolved is largely diminished because of the reutilization of butyrate for butanol synthesis.

Table 8.6. Fermentation balances of clostridia[a]

products	*C. butyricum*	*C. perfringens*	*C. acetobutylicum*
	(amounts formed in mol/100 mol glucose fermented)		
butyrate	76	34	4
acetate	42	60	14
lactate	—	33	—
CO_2	188	176	221
H_2	235	214	135
ethanol	—	26	7
butanol	—	—	56
acetone	—	—	22

[a]W. A. Wood, In: *The Bacteria*, I. C. Gunsalus and R. Y. Stanier (eds.). Academic Press, New York and London, 1961, vol. 2, pp. 59–149.

In complex fermentations like the one carried out by *C. acetobutylicum* it is difficult to judge whether the hydrogen balance is even or not. Therefore, the **oxidation reduction (O/R) balance** of complex fermentations is determined, as illustrated in Table 8.7. Arbitrarily the O/R value of formaldehyde and multiples thereof are taken as zero. Each 2 H in excess is expressed as −1, and a lack of 2H is expressed as +1. Ethanol has the sum formula C_2H_6O; addition of H_2O gives $C_2H_8O_2$. In comparison to $C_2H_4O_2$, 4H are in excess (O/R-value = −2). From CO_2 water is substracted; in C(−2H)O four hydrogens are missing as compared to CH_2O (O/R-value = +2).

Alternatively, such calculations can be done on the basis of the **number of available hydrogen** in the substrate and in the products. This

Table 8.7. Carbon recovery, O/R balance, and balance of available hydrogen of an acetone-butanol fermentation[a]

substrate and products	mol/ 100 mol substrate	mol carbon	O/R balance		balance of available hydrogen	
			O/R value	O/R value (mol/100 mol)	available H	available H (mol/100 mol)
glucose	100	600	0	—	24	2,400
butyrate	4	16	−2	−8	20	80
acetate	14	28	0	—	8	112
CO_2	221	221	+2	+442	0	—
H_2	135	—	−1	−135	2	270
ethanol	7	14	−2	−14	12	84
butanol	56	224	−4	−224	24	1,344
acetone	22	66	−2	−44	16	352
total		569		−425 +442		2,242

[a]Carbon recovered: $569/600 \times 100 = 94.8\%$; O/R balance: $442/425 = 1.04$; balance of available H: $2,400/2,242 = 1.07$. The latter two values are different because the carbon recovery affects the balance of available H but not the O/R balance (O/R value of glucose = 0).

number is determined by oxidizing the compound to CO_2 with water $(C_6H_{12}O_6 + 6H_2O \rightarrow \mathbf{24H} + 6CO_2; \; C_3H_6O + 5H_2O \rightarrow \mathbf{16H} + 3CO_2)$.

E. Not all saccharolytic clostridia form butyrate

Clostridia are always brought into close connection with the butyrate fermentation; and it is, therefore, important to stress that several saccharolytic clostridia do not produce butyrate at all. *C. sphenoides*, for instance, ferments glucose to ethanol, acetate, CO_2, and hydrogen. Furthermore, a few clostridial species (*C. thermoaceticum* and *C. formicoaceticum*) ferment hexose to almost three acetate; they will be discussed together with the methane bacteria.

F. The ethanol-acetate fermentation of *Clostridium kluyveri*

C. kluyveri carries out a very interesting fermentation. It produces butyrate, caproate, and hydrogen from ethanol and acetate. Ethanol alone is not fermented, but acetate can be replaced by propionate. The ratio in which butyrate and caproate are formed can vary; an increase of the ethanol concentration of the medium favors caproate formation. A typical fermentation balance is:

6 ethanol + 3 acetate $\longrightarrow$ 3 butyrate + 1 caproate + $2H_2$ + $4H_2O$ + H^+

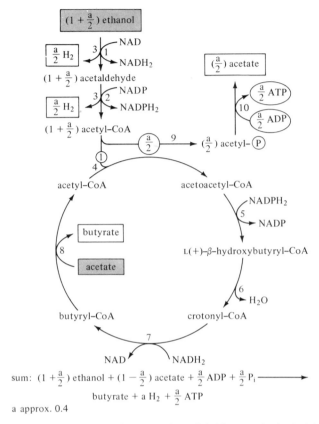

sum: $(1 + \frac{a}{2})$ ethanol $+ (1 - \frac{a}{2})$ acetate $+ \frac{a}{2}$ ADP $+ \frac{a}{2}$ P$_i$ ⟶

butyrate $+$ a H$_2$ $+ \frac{a}{2}$ ATP

a approx. 0.4

Figure 8.11. The ethanol-acetate fermentation of *C. kluyveri*. 1, alcohol dehydrogenase; 2, acetaldehyde dehydrogenase; 3, H$_2$-evolving enzyme system; 4, thiolase; 5, L(+)-β-hydroxybutyryl-CoA dehydrogenase (NADP-specific); 6, crotonase; 7, butyryl-CoA dehydrogenase; 8, CoA transferase; 9, phosphotransacetylase; 10, acetate kinase.

Approximately 0.3 mol of hydrogen is evolved per mol of ethanol fermented.

How *C. kluyveri* gains ATP for growth is difficult to see. The principle is that per mol of hydrogen evolved, 0.5 mol of acetyl-CoA is not required as H-acceptor as has been shown by Decker and collaborators. This acetyl-CoA serves to form ATP. The reactions involved are summarized in Figure 8.11. For the sake of simplicity only the formation of butyrate and H$_2$ is considered. Reactions involved in caproate formation are analogous (butyryl-CoA condenses with acetyl-CoA, 3-oxocaproyl-CoA is reduced, water is removed, and subsequent reduction yields caproyl-CoA).

Alcohol and acetaldehyde dehydrogenases are associated with one another and are particle bound. The first enzyme is NAD-specific and the second one reacts with either NAD and/or NADP. Hydrogen is evolved in the course of these reactions by a mechanism not fully understood. Presumably electrons are transferred from $NADH_2$ to hydrogenase via ferredoxin. A $NADH_2$-ferredoxin reductase, which requires acetyl-CoA for activity, has been demonstrated in cell extracts of *C. kluyveri*. $L(+)$-β-Hydroxybutyryl-CoA dehydrogenase is NADP-specific, which is unique among butyrate-forming organisms. It indicates a close coupling between acetaldehyde oxidation and acetoacetyl-CoA reduction.

In Figure 8.11 "a" has a value of about 0.4 so that approximately 0.4 mol H_2 are evolved per 1.2 mol ethanol. From the 1.2 mol of acetyl-CoA formed, 1 mol is required for $NADH_2$ and $NADPH_2$ oxidation in the butyric acid cycle; 0.2 mol is available for ATP formation. Thus, the ATP yield is 0.2/1.2 or 1 ATP per 6 ethanol.

G. Aspects of biosynthetic metabolism of *Clostridium kluyveri*

Although *C. kluyveri* grows on C_2-compounds, it does not contain a glyoxylate cycle for the synthesis of C_4-dicarboxylic acids. Instead, it takes advantage of the reversibility of the pyruvate-ferredoxin oxidoreductase reaction. Under the reducing conditions of the ethanol-acetate fermentation the reductive carboxylation of acetyl-CoA is feasible:

$$CH_3CO-SCoA + CO_2 + Fd \cdot H_2 \rightleftharpoons CH_3-CO-COOH + HSCoA + Fd$$

Pyruvate is trapped by formation of alanine or by further carboxylation to oxaloacetate by pyruvate carboxylase. The importance of carboxylation reactions in biosynthetic metabolism of *C. kluyveri* is underlined by the fact that *C. kluyveri* requires CO_2 for growth. About 30% of its cellular material is derived from CO_2. In agreement with the above-mentioned carboxylation reactions, the alanine carboxyl group and both carboxyl groups of aspartate originate quantitatively from CO_2. Since *C. kluyveri* contains also a pyruvate-formate lyase, the carboxyl group of pyruvate is also the precursor of formate, which is the starting material for C_1-units in biosynthesis.

Interestingly, *C. kluyveri* uses enzymes of the tricarboxylic acid cycle to synthesize glutamate; it contains citrate synthase, *cis*-aconitase, isocitrate dehydrogenase, and glutamate dehydrogenase, and so do many other obligate anaerobic bacteria. This emphasizes the dual function of the tricarboxylic acid portion of the cycle: provision of $NADH_2$ as part of the complete cycle (aerobes) and provision of α-oxoglutarate (aerobes and anaerobes).

The citrate synthase of *C. kluyveri* and of a few other anaerobes (*C. acidi-urici*, *C. cylindrosporum*, some sulfate-reducing bacteria) differs in two respects from all other citrate synthases. It has a specific requirement for Mn^{2+} whereas all other citrate synthases do not require metal ions at all. Moreover, it differs in its stereospecificity (Figure 8.12). Thus, a very small group of anaerobic bacteria contain a special enzyme, **re**-citrate synthase. All other organisms contain—as far as tested—the **si**-type synthase.

Figure 8.12. Formation of citrate by **si**-citrate synthase and by **re**-citrate synthase. Note that the carboxymethyl residue derived from acetyl-CoA comes into different positions. A, molecule turned 120° ; B, stereospecific removal of water by *cis*-aconitase allows distinction between the two carboxymethyl groups of radioactive citrate prepared. With ^{14}C-labeled acetyl-CoA, 4,5-^{14}C-*cis*-aconitate and 1,2-^{14}C-*cis*-aconitate are formed, respectively.

IV. Mixed Acid and Butanediol Fermentation

This type of fermentation is carried out by the enterobacteria. Organisms belonging to the genera *Escherichia*, *Salmonella*, and *Shigella* ferment sugars to lactic, acetic, succinic, and formic acids. In addition CO_2, H_2, and ethanol are formed. Species of the genera *Enterobacter*, *Serratia*, and *Erwinia* produce less acids but more gas (CO_2), ethanol, and above all large amounts of 2,3-butanediol. Two typical fermentation balances are given in Table 8.8, and the pathways leading to all these products are summarized in Figure 8.13.

Enterobacteria employ the Embden–Meyerhof pathway for hexose breakdown. The pathway leading to succinate branches off at phospho-enolpyruvate; all other end products are derived from pyruvate. Three enzyme systems act upon pyruvate and the amounts in which the fermentation products are formed depend very much on the activity of these enzyme systems. In the mixed acid fermentation large amounts of lactate are formed by the action of lactate dehydrogenase. Little lactate only is produced in the butanediol fermentation. The two other enzyme systems—pyruvate-formate lyase and α-acetolactate synthase deserve special attention.

Figure 8.13. Mixed acid (a) and butanediol (b) fermentation. 1, enzymes of the Embden–Meyerhof pathway; 2, lactate dehydrogenase; 3, pyruvate-formate lyase; 4, formate-hydrogen lyase; 5, acetaldehyde dehydrogenase; 6, alcohol dehydrogenase; 7, phosphotransacetylase; 8, acetate kinase; 9, PEP carboxylase; 10, malate dehydrogenase, fumarase, and fumarate reductase; 11, α-acetolactate synthase; 12, α-acetolactate decarboxylase; 13, 2,3-butanediol dehydrogenase.

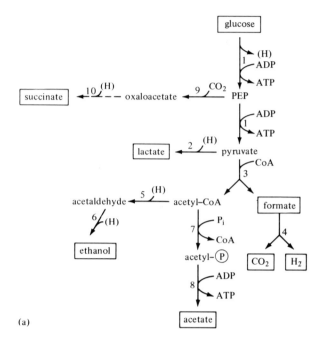

(a)

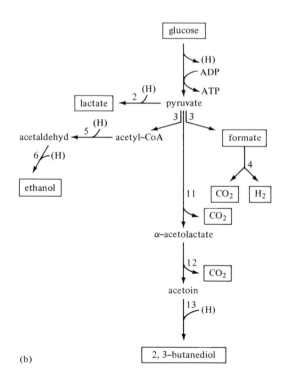

(b)

Table 8.8. Products formed in the mixed acid and butanediol fermentation[a]

product	*Escherichia coli*	*Enterobacter aerogenes*
	(mol formed/100 mol glucose fermented)	
formate	2.4	17.0
acetate	36.5	0.5
lactate	79.5	2.9
succinate	10.7	—
ethanol	49.8	69.5
2,3-butanediol	0.3	66.4
CO_2	88.0	172.0
hydrogen	75.0	35.4

[a]W. A. Wood, In: *The Bacteria*, I. C. Gunsalus and R. Y. Stanier (eds.). Academic Press, New York and London, 1961, vol. 2, pp. 59–149.

A. Pyruvate-formate lyase

The enterobacteria are able to synthesize two enzyme systems for pyruvate breakdown to acetyl-CoA. The pyruvate dehydrogenase multienzyme complex is involved in aerobic metabolism. Under anaerobic conditions it is no longer synthesized, and the enzyme still present is inhibited by $NADH_2$. Instead, the synthesis of pyruvate-formate lyase is induced under anaerobic conditions. The reaction catalyzed by this enzyme proceeds in two steps with an acetyl-enzyme as intermediate and formate and acetyl-CoA as products:

$$CH_3{-}CO{-}COOH + enzyme \rightleftharpoons CH_3{-}CO\text{-enzyme} + HCOOH$$
$$CH_3{-}CO\text{-enzyme} + CoASH \rightleftharpoons enzyme + CH_3{-}CO{-}SCoA$$

Pyruvate-formate lyase is irreversibly and rapidly inactivated under air so that it functions only in fermentative metabolism of the enterobacteria. Apparently, even under anaerobic conditions the active enzyme is not very stable; at low concentrations of pyruvate it changes over to an inactive form, which can again be reactivated. This reactivation requires the presence of four components: reduced flavodoxin, activating enzyme (enzyme II), S-adenosylmethionine, and pyruvate. The latter is not consumed in the activation reaction; thus, it functions as positive effector.

pyruvate-formate lyase   pyruvate-formate lyase
(inactive)   (active)
+   +

reduced flavodoxin $\xrightarrow[\text{(pyruvate)}]{\text{enzyme II}}$ flavodoxin
+   +

S-adenosylmethionine   methionine + 5′-deoxyadenosine

S-adenosylmethionine is reductively cleaved in this activation reaction. How and where the formate lyase actually is modified is not known.

The advantage of the pyruvate-formate lyase over the pyruvate dehydrogenase complex in fermentative metabolism is apparent: the formation of acetyl-CoA is not accompanied by the reduction of NAD.

B. α-Acetolactate synthase

The third enzyme system acting upon pyruvate is α-acetolactate synthase. This enzyme is also involved in 2,3-butanediol formation by bacilli (see Figure 6.24). It contains thiamin pyrophosphate. First, enzyme-bound hydroxyethyl-thiamin pyrophosphate and CO_2 are formed from pyruvate. Active acetaldehyde is then transferred to a second molecule of pyruvate:

$$CH_3-CO-COOH + (H)TPP-E \longrightarrow CH_3-\underset{\underset{OH}{|}}{CH}-TPP-E + CO_2$$

$$\begin{array}{c} CH_3-\underset{\underset{OH}{|}}{CH}-TPP-E \\ + \\ CH_3-CO-COOH \end{array} \longrightarrow \begin{array}{c} CH_3-CO \\ | \\ CH_3-\underset{\underset{OH}{|}}{C}-COOH \end{array} + (H)TPP-E$$

This synthase is formed and is active under slightly acidic conditions, and it is referred to as the pH 6-enzyme. Thus, a decrease of pH in the environment of *Enterobacter aerogenes* leads to an increase of 2,3-butanediol formation. Consequently less acids can be produced from pyruvate.

The pH 6-enzyme is distinct from the anabolic α-acetolactate synthase, which is involved in valine synthesis. This enzyme is most active at pH 8 (pH 8-enzyme) and is subject to feedback inhibition by L-valine.

C. Formate-hydrogen lyase

Species belonging to the genera *Shigella* and *Erwinia* do not contain formate-hydrogen lyase; they produce considerable amounts of formate. *Escherichia coli* and *Enterobacter aerogenes* contain this enzyme when grown on sugars under anaerobic conditions, and most of the formate is cleaved into CO_2 and H_2.

Formate-hydrogen lyase consists of several components. As is apparent from Figure 8.14(a), formate is oxidized to CO_2 by formate dehydrogenase. The electrons are passed over the carriers X_1 and X_2 to hydrogenase, at which H_2 is evolved.

The nature of the carriers X_1 and X_2 is not known; none of them is a cytochrome. Formate-hydrogen lyase resides in the enterobacteria as a membrane-bound multienzyme complex. It is of particular interest that the formation of active formate dehydrogenase component requires selenite and molybdate. Se and Mo are present in the enzyme.

The formate-hydrogen lyase complex is induced by formate; its synthesis

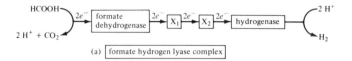

(a) [formate hydrogen lyase complex]

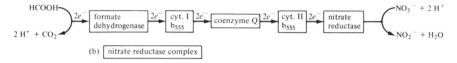

(b) [nitrate reductase complex]

Figure 8.14. The components of the formate-hydrogen lyase complex (a) and the nitrate reductase complex (b) in *E. coli*. X_1 and X_2 are carriers of unknown structure.

is repressed by oxygen and also by nitrate under anaerobic conditions. Under the latter conditions formate serves as electron donor in the dissimilatory reduction of nitrate to nitrite (nitrate/nitrite respiration; see Chapter 5, Section VI). As shown in Figure 8.14(b) a multienzyme complex distinct from the formate-hydrogen lyase complex is responsible for electron transfer from formate to nitrate. It includes a formate dehydrogenase, which contains also Se and Mo but differs in kinetic properties from the corresponding enzyme of the hydrogen lyase complex. Furthermore, two carrier proteins with cytochrome b_{555} as prosthetic group, coenzyme Q, and nitrate reductase are present. The latter also is a molybdoprotein.

D. ATP yield

The ATP yield of the mixed acid and butanediol fermentation lies between 2 and 2.5 moles ATP/mol glucose. In addition to the two ATP gained in the Embden–Meyerhof pathway some ATP is formed in acetate formation by acetate kinase.

V. Propionate and Succinate Fermentation

Propionate is a major end product of fermentations carried out by a variety of anaerobic bacteria. Many of them ferment glucose to propionate, acetate, and CO_2:

$$1.5 \text{ glucose} \longrightarrow 2 \text{ propionate} + \text{acetate} + CO_2$$

A preferred substrate of propionate-forming bacteria is lactate, so that these organisms can grow with the major end product of the lactate fermentation:

$$3 \text{ lactate} \longrightarrow 2 \text{ propionate} + \text{acetate} + CO_2$$

There are two pathways for propionate formation; in the acrylate pathway lactate is reduced stepwise to propionate; in the succinate-propionate pathway lactate is converted to propionate via pyruvate and succinate.

A. The acrylate pathway

This pathway seems to occur only in a few microorganisms, e.g., in *Clostridium propionicum* and in *Megasphaera (Peptostreptococcus) elsdenii*. It is shown in Figure 8.15. L-, D-, or DL-Lactate may serve as substrate; a racemase is present which interconverts the enantiomers. L-Lactate is converted to L-lactyl-CoA in a CoA transferase reaction. By reactions not yet established in detail acrylyl-CoA is formed. It is reduced to propionyl-CoA, and propionate is produced by the above-mentioned CoA transferase.

The H-donor for acrylyl-CoA reduction is reduced electron-transferring flavoprotein. It is formed from D-lactate and from reduced ferredoxin (or flavodoxin). The ATP yield of this fermentation is 1 mol/3 mol of lactate.

C. propionicum also ferments alanine and acrylate to propionate.

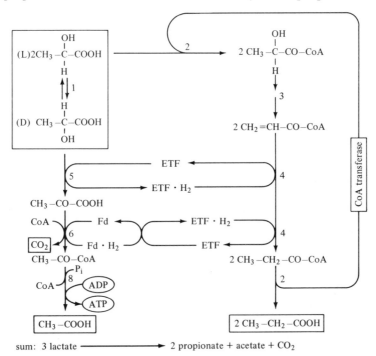

Figure 8.15. Formation of propionate, acetate, and CO_2 from DL-lactate by *Megasphaera elsdenii* and *Clostridium propionicum*. 1, lactate racemase; 2, CoA transferase; 3, reaction not established; 4, dehydrogenase, which employs reduced electron-transferring flavoprotein ($ETF \cdot H_2$) as H-donor; 5, D-lactate dehydrogenase; 6, pyruvate-ferredoxin oxidoreductase; 7, transhydrogenase; 8, phosphotransacetylase + acetate kinase.

B. The succinate-propionate pathway

This pathway is employed by most propionate-producing organisms. Succinate is an intermediate but is also produced as end product in small or large amounts. On the other hand, organisms using the acrylate pathway do not excrete significant amounts of succinate.

The establishment of the succinate-propionate pathway was a rather difficult task. As is shown in Figure 8.16 several enzymes are involved. First, lactate is oxidized to pyruvate in a reaction requiring a flavoprotein as H-acceptor. Oxaloacetate is then formed in a transcarboxylation reaction with (S)-methylmalonyl-CoA as CO_2-donor and biotin as CO_2-carrier. The action of malate dehydrogenase and fumarase yields fumarate, which is reduced to succinate by fumarate reductase. This reduction reaction is coupled to ATP formation by electron transport phosphorylation. Succinyl-CoA is then formed in a CoA transferase reaction and the rearrangement as catalyzed by the coenzyme B_{12}-containing methylmalonyl-CoA mutase leads

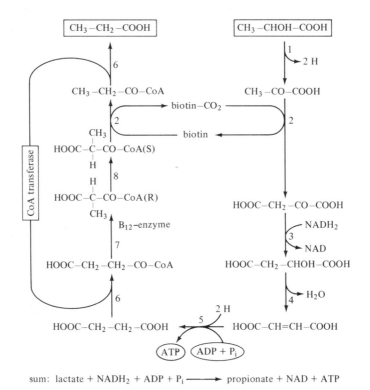

sum: lactate + $NADH_2$ + ADP + P_i ⟶ propionate + NAD + ATP

Figure 8.16. Fermentation of lactate via the succinate-propionate pathway by propionibacteria. 1, lactate dehydrogenase (the H-acceptor is probably a flavoprotein); 2, (S)-methylmalonyl-CoA-pyruvate transcarboxylase; 3, malate dehydrogenase; 4, fumarase; 5, fumarate reductase; 6, CoA transferase; 7, (R)-methylmalonyl-CoA mutase; 8, methylmalonyl-CoA racemase.

to (R)-methylmalonyl-CoA, which is not a substrate for the transcarboxylase. Rather, the (S)-enantiomer is formed by a specific racemase. Then trans-carboxylation yields propionyl-CoA and CoA transfer to succinate finally yields propionate.

One $NADH_2$ is consumed in propionate formation from lactate; it comes from lactate oxidation to acetate according to the overall fermentation equation given above.

Besides the transcarboxylase—a biotin-containing enzyme with a high molecular weight (approximately 800,000 daltons) and a very complex quarternary structure—two enzymes of the succinate-propionate pathway deserve special attention: methylmalonyl-CoA mutase and fumarate reductase.

C. Methylmalonyl-CoA mutase and other coenzyme B_{12} – dependent rearrangement reactions

Rearrangements of this type were discovered by Barker and collaborators when they investigated the fermentation of glutamate by *Clostridium tetanomorphum*. As is apparent from Figure 8.17 glutamate and succinyl-CoA are rearranged in analogous reactions to yield β-methylaspartate and methylmalonyl-CoA, respectively. The principle of these reactions is that a substituent group is moved between two adjacent positions of the carbon skeleton while a hydrogen is moved in the opposite direction. Not only carbon-carbon bonds are rearranged in coenzyme B_{12}-dependent reactions. A number of dehydratases, deaminases, and amino mutases are also B_{12}-enzymes and catalyze analogous reactions [Figure 8.17(c) to (e)]. Glycerol dehydrase, which is present in some lactobacilli, converts glycerol into β-hydroxypropionaldehyde. Ethanolamine deaminase is present in choline-fermenting clostridia and β-lysine mutase is the second enzyme in clostridial L-lysine fermentation (see Chapter 8, Section VIII).

Coenzyme B_{12} is not identical with **vitamin B_{12}**. The latter is a corrin ring system with cobalt^{2+} as central metal atom and 5,6-dimethylbenzimidazole ribonucleotide as characteristic component (Figure 8.18). The sixth co-ordination position of Co^{2+} is occupied by hydroxyl or cyanide (hydroxy- or cyanocobalamin). Coenzyme B_{12} contains in addition a 5'-deoxyadenosyl group, which is covalently bound to cobalt replacing cyanide (5'-deoxy-adenosylcobalamin). In the rearrangement reactions the hydrogen is transferred from the substrate to coenzyme B_{12} (to the C-5' methylene group) and from there to the product.

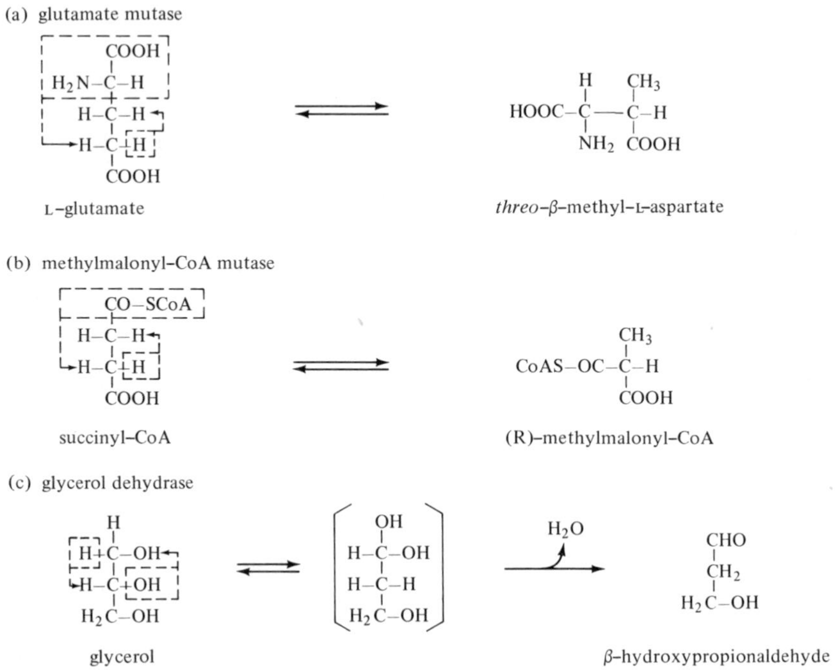

(a) glutamate mutase

L-glutamate   threo-β-methyl-L-aspartate

(b) methylmalonyl-CoA mutase

succinyl-CoA   (R)-methylmalonyl-CoA

(c) glycerol dehydrase

glycerol   β-hydroxypropionaldehyde

Figure 8.17. Coenzyme B_{12}-dependent rearrangement reactions.

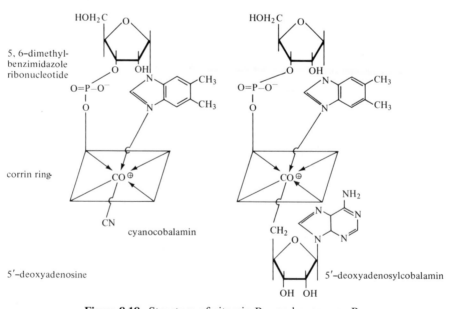

Figure 8.18. Structure of vitamin B_{12} and coenzyme B_{12}.

(d) ethanolamine deaminase

$$H_2C-OH \cdots H_2C-NH_2 \rightleftharpoons \left(\begin{array}{c} NH_2 \\ H-C-OH \\ CH_3 \end{array} \right) \xrightarrow{NH_3} \begin{array}{c} CHO \\ CH_3 \end{array}$$

ethanolamine   acetaldehyde

(e) β–lysine–5, 6–aminomutase

$$\begin{array}{c} COOH \\ CH_2 \\ H_2N-C-H \\ CH_2 \\ H-C-H \\ H_2C-NH_2 \end{array} \rightleftharpoons \begin{array}{c} COOH \\ CH_2 \\ H_2N-C-H \\ CH_2 \\ H_2N-C-H \\ CH_3 \end{array}$$

β–lysine   3, 5–diaminohexanoate

Figure 8.17.   Continued.

D. Fumarate reductase

The reduction of fumarate to succinate is the only known process by which strict anaerobes can gain ATP by electron transport phosphorylation. That ATP is formed in this reaction is indicated by the high growth yields of propionate-succinate-producing organisms. Moreover, an ATP synthesis coupled to the reduction of fumarate could be demonstrated in cell-free systems of some organisms, first with extracts of *Desulfovibrio gigas* by Peck, Le Gall, and Barton. Fumarate reductase is membrane-bound, and it is associated with the electron carrier menaquinone (for the formula see Figure 2.8). In several but not in all microorganisms a cytochrome b is also part of the fumarate reductase system. *Bacteroides fragilis* grows only slowly in the absence of hemin (precursor in cytochrome synthesis) and ferments glucose to lactate, acetate, and fumarate. In the presence of hemin, however, a cytochrome b is formed and glucose is rapidly fermented to propionate and succinate. In this microorganism cytochrome b is an essential component of the fumarate reductase system. *Bacteroides amylophilus*, on the other hand, forms succinate without the involvement of a cytochrome.

Electron flow from H_2 and $NADH_2$ to fumarate is summarized in Figure 8.19. The redox potential of fumarate/succinate is $E_0' = +0.03$ V. Thus the potential span between $NADH_2/NAD$ ($E_0' = -0.32$ V) and fumarate/succinate is large enough to allow ATP synthesis by electron transport phosphorylation.

The fumarate reductase system is present not only in the classical propionate-forming microorganisms, the propionibacteria, but also in several enterobacteria and in various species of the genera *Bacteroides*, *Veillonella*, *Peptostreptococcus*, *Ruminococcus*, *Succinivibrio*, and *Selenomonas*. Even a clostridium (*C. formicoaceticum*) contains fumarate reductase. Some of these bacteria form only succinate but no propionate. Finally, a very interesting fermentation should be mentioned: the fermentation of fumarate to

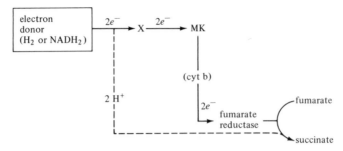

Figure 8.19. Electron flow from H_2 and $NADH_2$ as H-donors to fumarate. X, carrier of unknown structure; MK, menaquinone.

succinate with H_2 or formate as hydrogen donors:

$$\text{fumarate} + H_2 \longrightarrow \text{succinate}$$
$$\text{fumarate} + \text{formate} \longrightarrow \text{succinate} + CO_2$$

Organisms like *Vibrio succinogenes*, almost all enterobacteria, and *Desulfovibrio gigas* can grow on the basis of this fermentation. This again demonstrates that ATP synthesis is coupled to the reduction of fumarate to succinate.

E. PEP carboxytransphosphorylase of propionibacteria

Since propionibacteria produce succinate in addition to propionate, the transcarboxylase alone cannot be responsible for the formation of C_4-dicarboxylic acids; an enzyme system must be present in these microorganisms that catalyzes the synthesis of C_4-dicarboxylic acids from a C_3-compound and CO_2. It was discovered by Wood and Werkman in 1936. Since then carboxylating reactions which yield oxaloacetate are called **Wood–Werkman reactions**. We have already discussed the role of pyruvate carboxylase and PEP carboxylase as C_3-carboxylating enzymes. In some *Bacteroides* species the energy-conserving PEP carboxykinase reaction is involved in oxaloacetate formation (this enzyme functions normally in the direction of PEP synthesis; see Table 5.4):

$$PEP + CO_2 + ADP \xrightleftharpoons{\text{PEP carboxykinase}} \text{oxaloacetate} + ATP$$

Propionibacteria contain an enzyme that catalyzes an analogous reaction:

$$PEP + CO_2 + P_i \xrightarrow{\text{PEP carboxytransphosphorylase}} \text{oxaloacetate} + PP_i$$

Here, the phosphoryl group is transferred to inorganic phosphate to form pyrophosphate. Thus, propionibacteria contain two enzyme systems that yield pyrophosphate: PEP carboxytransphosphorylase and pyruvate-phosphate dikinase (see Table 5.4). The pyrophosphate formed is used to phosphorylate fructose-6-phosphate to fructose-1,6-bisphosphate and serine to phosphoserine.

VI. Methane and Acetate Fermentation

Methane is the most reduced organic compound and its formation is the terminal step in the food chain of fermentative microorganisms. Methanogenic bacteria are common in anaerobic environments where organic matter is being decomposed. They are extremely oxygen-sensitive, and, until recently, their isolation and growth in pure culture has been very difficult. Now methods are available (Hungate technique) to grow these microorganisms on agar and in liquid culture.

Methane bacteria are not able to utilize complex organic compounds; they grow with substrates such as $CO_2 + H_2$, formate, methanol, and acetate:

$$CO_2 + 4H_2 \longrightarrow CH_4 + 2H_2O$$
$$4HCOOH \longrightarrow CH_4 + 3CO_2 + 2H_2O$$
$$4CH_3OH \longrightarrow 3CH_4 + CO_2 + 2H_2O$$
$$CH_3COOH \longrightarrow CH_4 + CO_2$$

The most important substrates in nature are $CO_2 + H_2$ and acetate.

A. Methane formation from CO_2 and H_2

All methanogenic bacteria isolated thus far can grow with CO_2 and H_2, and the reactions involved in methane formation have been studied using *Methanobacterium ruminantium*, *M. thermoautotrophicum*, and *Methanobacterium MoH*. The reduction of CO_2 to CH_4 proceeds stepwise but the intermediates (formate, formaldehyde, and methanol) remain firmly bound to carriers that are as yet only partly known. One important carrier was recently discovered by Wolfe and collaborators, coenzyme M (2-mercaptoethanesulfonic acid):

coenzyme M: $HS-CH_2-CH_2-SO_3H$
methylcoenzyme M: $CH_3-S-CH_2-CH_2-SO_3H$

methylcobalamin

Methylcoenzyme M is probably the direct precursor of methane. Furthermore, methylcobalamin serves as an intermediate.

The carrier at which CO_2 is reduced to the oxidation level of formate and formaldehyde is not known. Tetrahydrofolate seems not to be involved in methane formation.

A scheme based on Barker's general scheme of methane formation from CO_2 is given in Figure 8.20. CO_2 is reduced to carrier-methyl and the methyl group is then transferred via B_{12} or directly to coenzyme M. CH_4 is formed in a final reductive step.

Methanogenic bacteria contain a fluorescent compound, F_{420}, as hydrogen carrier. F_{420} has not been detected in other microorganisms and because of the strong fluorescence of this compound methanogenic bacteria can be easily recognized as such in a fluorescence microscope.

It is apparent from Figure 8.20 that the formation of methane from CO_2 and H_2 cannot be coupled to ATP synthesis by substrate-level phosphorylation. Thus, it must be postulated that methanogenic bacteria gain ATP by

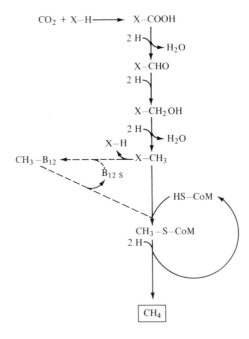

Figure 8.20. Scheme for the reduction of CO_2 to methane. X, carrier of unknown structure; B_{12s}, hydridocobalamin; HS-CoM, coenzyme M.

electron transport phosphorylation. The free energy change for the reduction of bicarbonate to methane is as follows:

$$\Delta G_0'$$

$HCO_3^- + H_2$	$\longrightarrow HCOO^- + H_2O$	$-\ 0.3\ kcal\ (-1.3kJ)$
$HCOO^- + H^+ + H_2$	$\longrightarrow CH_2O + H_2O$	$+\ 5.5\ kcal\ (+23.0kJ)$
$CH_2O + H_2$	$\longrightarrow CH_3OH$	$-10.7\ kcal\ (-44.8kJ)$
$CH_3OH + H_2$	$\longrightarrow CH_4 + H_2O$	$-26.9\ kcal\ (-112.5kJ)$

$$HCO_3^- + H^+ + 4H_2 \longrightarrow CH_4 + 3H_2O \qquad -32.4\ kcal\ (-135.6kJ)$$

The free energy change of the overall process is very negative but the values for the individual reduction steps indicate that the major part of this change is contributed by the reduction of methanol to methane. The redox potential of the methanol/methane couple is $+0.170$ V, and as in the reduction of fumarate with H_2, electron flow from hydrogen to carrier-bound methanol may be coupled to ATP synthesis. This is indicated in Figure 8.21. It should be emphasized that experimental evidence for such a mechanism of ATP synthesis is still lacking.

Since methanogenic bacteria grow with CO_2 and H_2 they are chemolithotrophic bacteria. However, they have so very much in common with other fermentative organisms that it seems appropriate to place them in this chapter. Moreover, methanogenic bacteria do not seem to possess the reductive pentose phosphate cycle as the chemolithotrophs usually do (see Chapter 9). The mechanisms employed by methanogenic bacteria for synthesis of cellular material from CO_2 are not yet known.

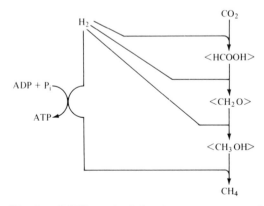

Figure 8.21. Possible site of ATP synthesis by electron transport phosphorylation in methanogenic bacteria.

B. Interspecies hydrogen transfer

Methane formation from CO_2 and H_2 is a very effective mechanism for trapping any hydrogen evolved in a fermentative process. In the rumen and in anaerobic digesters in which CH_4 production from $CO_2 + H_2$ is abundant, the partial pressure of hydrogen is kept as low as 10^{-4} atm by the methanogenic bacteria. This is very important. At high partial pressure, hydrogen can only be formed from pyruvate (pyruvate-ferredoxin oxidoreductase + hydrogenase) and from formate (formate-hydrogen lyase); at low partial pressures hydrogen evolution from $NADH_2$ ($NADH_2$-ferredoxin reductase + hydrogenase) and other H-carriers becomes possible. Figure 8.22 illustrates that below a partial pressure of H_2 of 10^{-3} atm the free energy change of H_2 evolution from $NADH_2$ is negative. Thus, in mixed cultures where p_{H_2}

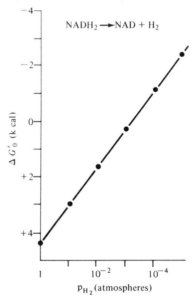

Figure 8.22. The free energy change for the oxidation of $NADH_2$ to NAD and H_2 as dependent on the partial pressure of hydrogen.

is kept low by the action of methanogenic bacteria, other fermentative processes are shifted toward production of more hydrogen and less reduced organic compounds. Two examples are given in Figure 8.23. The so-called S organism is able to oxidize ethanol to acetate and H_2. This is thermodynamically possible only at low p_{H_2}. Consequently the S organism can only grow in the presence of methanogenic bacteria. The mixed culture of the S organism and *Methanobacterium* MoH became known as *Methanobacterium omelianskii*.

$$2CH_3\text{---}CH_2OH + 2H_2O \xrightarrow{\text{S organism}} 2CH_3\text{---}COOH + 4H_2$$
$$4H_2 + CO_2 \xrightarrow{\textit{Methanobacterium}\ \text{MoH}} CH_4 + 2H_2O$$

$$2CH_3\text{---}CH_2OH + CO_2 \xrightarrow{\textit{"Methanobacterium omelianskii"}} 2CH_3\text{---}COOH + CH_4$$

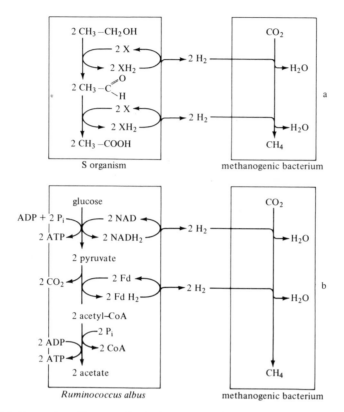

Figure 8.23. Examples of interspecies hydrogen transfer. At low p_{H_2} ethanol is oxidized by the S organism to acetate and H_2 and glucose is oxidized to acetate $+ CO_2 + H_2$. The p_{H_2} is kept low by the action of methanogenic bacteria.

Ruminococcus albus, which has been studied by Wolin and collaborators, ferments glucose to ethanol, acetate, and hydrogen:

$$\text{glucose} \longrightarrow \text{acetate} + \text{ethanol} + 2CO_2 + 2H_2$$

In the presence of methanogenic bacteria only acetate and CO_2 are formed. This is of benefit for both types of organisms. *R. albus* gains $4ATP/glucose$ as compared to $3ATP/glucose$ in pure culture; the methanogenic bacteria require, of course, H_2 and CO_2 for energy production and growth.

It should be emphasized that food chains of the type discussed here are of great importance in nature.

C. Methane formation from formate, methanol, and acetate

Formate is a good substrate for many methanogenic bacteria; it is first converted to CO_2 and hydrogen. It is thus not a direct precursor of methane:

$$4HCOOH \xrightarrow{\text{formate dehydrogenase}} 4CO_2 + 4H_2$$
$$CO_2 + 4H_2 \xrightarrow{\hspace{3cm}} CH_4 + 2H_2O$$

$$\text{sum:} \quad 4HCOOH \xrightarrow{\hspace{3cm}} CH_4 + 3CO_2 + 2H_2O$$

Methanol is a good substrate for *Methanosarcina barkeri*. It is a direct precursor of methane. *Methanosarcina barkeri* contains an enzyme system by which methanol and hydridocobalamin are converted to methyl-B_{12}. The latter then gives rise to the formation of methane:

$$CH_3OH + B_{12s} \xrightarrow[\text{and cofactors}]{\text{several proteins}} CH_3{-}B_{12} + H_2O$$

The reducing power for the formation of methane from $CH_3{-}B_{12}$ via $CH_3{-}CoM$ is furnished by the oxidation of methanol to CO_2, and the stoichiometry of this fermentation is as follows:

$$CH_3OH + H_2O \longrightarrow CO_2 + 6H$$
$$3CH_3OH + 6H \longrightarrow 3CH_4 + 3H_2O$$

$$\text{sum:} \quad 4CH_3OH \longrightarrow 3CH_4 + CO_2 + 2H_2O$$

The oxidation of methanol to CO_2 proceeds via methyltetrahydrofolate (methyl-H_4 folate), methylene-H_4 folate, methenyl-H_4 folate, and formyl-H_4 folate; from the latter compound formate is produced by **formyl-H_4 folate synthetase**:

$$ADP + P_i + \text{formyl-}H_4 \text{ folate} \rightleftharpoons \text{formate} + H_4 \text{ folate} + ATP$$

Thus, in the course of methanol oxidation, ATP is synthesized by substrate-level phosphorylation. Since the oxidation of 1 methanol is coupled to the reduction of 3 methanol to methane, 1 ATP/4 methanol is gained by substrate-level phosphorylation.

The final oxidation step is catalyzed by a formate dehydrogenase, which transfers hydrogen to the carrier F_{420}:

$$HCOOH + F_{420} \rightleftharpoons CO_2 + \text{reduced } F_{420}$$

Acetate is the most important methanogenic substrate in lake sediments. It is fermented by *Methanosarcina barkeri*, *Methanospirillum hungatii*, and some other methanogenic bacteria to methane and CO_2.

$$CH_3{-}COOH \longrightarrow CH_4 + CO_2$$

Barker and Stadtman have shown that the methyl carbon of acetate together with its attached hydrogens is converted to methane and that CO_2

originates from the carboxyl group of acetate. The mechanism of this fermentation remains a mystery.

D. Acetate fermentation

In 1936 Wieringa described *Clostridium aceticum*, which ferments hydrogen and carbon dioxide to acetate:

$$4H_2 + 2CO_2 \longrightarrow CH_3-COOH + 2H_2O$$

It is apparent that this fermentation is closely related to the methane fermentation. The free energy change is $\Delta G_0' = -25.6$ kcal (107.1 kJ), about 28 kJ less than in methane formation from H_2 and CO_2. Unfortunately, *C. aceticum* was later lost and details about its metabolism are not known. However, recently the isolation of a nonsporeformer (*Acetobacterium woodii*) carrying out the same type of fermentation has been reported by Wolfe and collaborators.

Two other clostridial species deserve attention in this connection: *C. thermoaceticum* and *C. formicoaceticum*. These microorganisms ferment hexoses almost completely to acetate. They are unable to evolve hydrogen and to form reduced compounds, such as lactate, ethanol, or butyrate, and their fermentative metabolism can be understood on the basis of the following equations:

$$C_6H_{12}O_6 + H_2O \longrightarrow CH_3-CO-COOH + CH_3COOH + CO_2 + 6H$$
$$CO_2 + 6H + HX \longrightarrow CH_3-X + 2H_2O$$
$$CH_3-CO-COOH + CH_3-X + H_2O \longrightarrow 2CH_3-COOH + H-X$$

$$\overline{C_6H_{12}O_6 \qquad\qquad \longrightarrow 3CH_3-COOH}$$

First, the hexose is degraded to pyruvate via the Embden–Meyerhof pathway. One of the pyruvate molecules is degraded further to acetate. The reducing power (6H) accumulated in the form of reduced coenzymes is used to reduce CO_2 to a carrier-bound methyl group (CH_3-X). Finally, pyruvate and CH_3-X react in a transcarboxylase reaction to give two acetates. The whole pathway is depicted in Figure 8.24.

CO_2 is reduced to formate by formate dehydrogenase. The enzyme contains selenium and tungsten. Thus it is different from formate dehydrogenase of *E. coli*, where molybdate cannot be replaced by tungstate.

Formate is then converted to formyl-H_4 folate by the corresponding synthetase. Reduction reactions lead to methyl-H_4 folate. Following transfer of the methyl group to a B_{12}-containing protein, a transcarboxylase forms two molecules of acetate from pyruvate and from the B_{12}-linked methyl group.

C. thermoaceticum and *C. formicoaceticum* ferment 1 mol of hexose to about 2.8 mol of acetate. *C. formicoaceticum* excretes some formate in

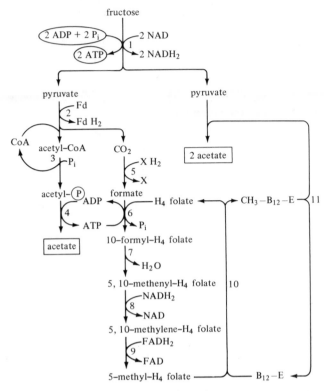

Figure 8.24. Acetate fermentation of *Clostridium formicoaceticum*. 1, degradation of fructose via the Embden–Meyerhof pathway; 2, pyruvate-ferredoxin oxidoreductase; 3, phosphotransacetylase; 4, acetate kinase; 5, formate dehydrogenase (electron acceptor unknown; in *C. thermoaceticum*, NADP); 6, formyl-H_4 folate synthetase; 7, methenyl-H_4 folate cyclohydrolase; 8, methylene-H_4 folate dehydrogenase; 9, methylene-H_4 folate reductase; 10, methyl transferase; 11, pyruvate-methyl-B_{12} transcarboxylase.

addition. The organisms differ in that the first is a thermophile and the second a mesophile. *C. formicoaceticum* is unable to ferment glucose but grows rapidly with fructose.

VII. Sulfide Fermentation (Desulfurication)

Most microorganisms use sulfate as the principal sulfur source and contain enzyme systems for the reduction of sulfate to sulfide. This process of assimilatory sulfate reduction has been discussed in Chapter 3 (see Figure 3.3). In sulfide fermentation, sulfate is used as terminal electron acceptor, and the hydrogen sulfide formed is excreted. This process is therefore called **dissimilatory sulfate reduction**; it is carried out only by strictly anaerobic bacteria of the genera *Desulfovibrio*, *Desulfomonas*, and *Desulfotomaculum*.

Sulfate-reducing bacteria utilize end products of other fermentations such as lactate, malate, and ethanol as H-donors: these compounds are oxidized to acetate (and CO_2). The reducing power generated is then employed for reduction of sulfate to sulfide. The preferred substrate of these microorganisms is lactate.

A. Fermentation of lactate and sulfate

This fermentation can be summarized as follows:

$$2CH_3-CHOH-COOH + 2H_2O \longrightarrow 2CH_3-COOH + 2CO_2 + 8H$$
$$SO_4^{2-} + 8H \longrightarrow S^{2-} + 4H_2O$$

Lactate oxidation to acetate proceeds via pyruvate and acetyl-CoA. A flavoprotein probably functions as H-acceptor in the first oxidation step, and pyruvate-ferredoxin oxidoreductase is involved in acetyl-CoA formation. The final acetate production is coupled to ATP synthesis.

The 8H generated in lactate oxidation are used to reduce sulfate to sulfide in four steps. Prior to the first reduction step, sulfate is activated by conversion to APS (Figure 8.25), and the reduction products are sulfite and AMP. The further phosphorylated compound PAPS, which is an intermediate in assimilatory sulfate reduction (see Figure 3.3), is not involved here. APS reductase contains FAD and iron-sulfur. The reaction is reversible; *in vitro* APS reduction is observed with reduced methyl viologen as H-donor and APS formation from sulfite and AMP with ferricyanide as H-acceptor.

The reduction of sulfite by *Desulfovibrio* species involves a recycling sulfite pool, and the first reduction product is trithionate. The sulfite reductase, which carries out this reaction, is identical with the green porphyrin-containing protein—**desulfoviridin**—discovered in sulfate-reducing bacteria by Postgate in 1956. Desulfoviridin is present in most *Desulfovibrio* species in large quantities but *Desulfotomaculum* species are devoid of this compound. They employ a sulfite reductase that exhibits an absorption spectrum different from that of desulfoviridin and that seems to reduce sulfite all the way to sulfide. The final two reduction steps in Figure 8.25 are catalyzed by trithionate reductase and by thiosulfate reductase.

It then can be concluded that two pathways of sulfite reduction exist in sulfate-reducing bacteria. One pathway involves a recycling sulfite pool, desulfoviridin, trithionate, and thiosulfate reductase, whereas the other one involves one enzyme system only.

Figure 8.25 does not give any details with regard to the electron carriers involved in the four reduction steps. In addition to ferredoxin, *Desulfovibrio* species contain rubredoxin, menaquinone, and cytochromes, most noteworthy being cytochrome c_3. All these carriers are involved in the hydrogen metabolism of *Desulfovibrio* but their exact reaction sites are not known.

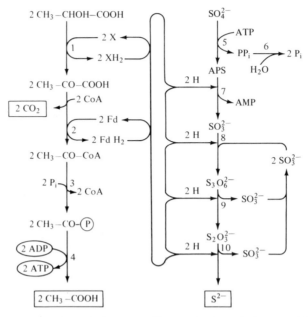

Figure 8.25. Pathway of dissimilatory sulfate reduction. 1, lactate dehydrogenase, H-acceptor not known; 2, pyruvate-ferredoxin oxidoreductase; 3, phosphotrans-acetylase; 4, acetate kinase; 5, ATP sulfurylase; 6, pyrophosphatase; 7, APS reductase; 8, sulfite reductase (desulfoviridin); 9, trithionate reductase; 10, thiosulfate reductase. The electron donors of the reductases are not known. APS, adenosine-5'-phospho-sulfate.

Desulfotomaculum species are devoid of cytochrome c_3 and contain cyto-chromes of the b type.

It is evident from Figure 8.25 that sulfate reducers gain two ATP in the oxidation of lactate to acetate. On the other hand, two high-energy bonds have to be invested in the reduction of sulfate to sulfite ($ATP \rightarrow AMP + 2P_i$). Thus, the overall yield of ATP gained by substrate-level phosphorylation is zero. Since many sulfate-reducing bacteria can utilize H_2 and form sulfide from sulfate + H_2, it follows that the process of sulfate reduction does not require an input of ATP, and it must be concluded that at least two ATP are formed by electron transport phosphorylation during the reduction of one sulfate to sulfide. The ATP yield might be even higher.

B. Growth with other substrates

Practically all sulfate-reducers are able to grow with ethanol + sulfate. Like *Clostridium kluyveri* these microorganisms then employ pyruvate-ferredoxin oxidoreductase for pyruvate synthesis and require the presence of bicarbonate in the growth medium. Malate + sulfate are fermented by *Desulfovibrio desulfuricans*, *D. salexigens*, and *D. africanus* and formate+sulfate by *Desulfotomaculum ruminis*. The latter and *D. desulfuricans* ferment pyruvate in the absence of sulfate to acetate, H_2, and CO_2.

C. Desulfuromonas acetoxidans

Recently, Pfennig and collaborators isolated a bacterium that grows with acetate and elemental sulfur; acetate is oxidized to CO_2 and the reducing power generated is used to reduce sulfur to sulfide:

$$CH_3-COOH + 2H_2O \longrightarrow 2CO_2 + 8H \qquad \Delta G_0' \doteq +25.5 \text{ kcal}(+106.7 \text{ kJ})$$
$$4S^0 + 8H \longrightarrow 4H_2S \qquad \Delta G_0' = -31.2 \text{ kcal}(-130.5 \text{ kJ})$$

$$CH_3-COOH + 2H_2O + 4S^0 \longrightarrow 2CO_2 + 4H_2S \quad \Delta G_0' = -5.7 \text{ kcal}(-23.8 \text{ kJ})$$

Desulfuromonas is devoid of normal fermentative activities and does not grow with pyruvate and with sugars. The mechanisms of acetate oxidation and ATP formation are not known.

VIII. Fermentation of Nitrogenous Compounds

Sugars and organic acids are not the only substrates for anaerobes. Amino acids (formed from proteins by extracellular proteases) and purine and pyrimidine bases are fermented by a variety of microorganisms.

A. Single amino acids

A number of single amino acids can serve as energy and carbon source for anaerobes. **Alanine** is fermented by *Clostridium propionicum* via the acrylate pathway (see Figure 8.15). *Peptococcus anaerobius* (*Diplococcus glycinophilus*) ferments **glycine** according to the following equation:

$$4H_2N-CH_2-COOH + 2H_2O \longrightarrow 4NH_3 + 2CO_2 + 3CH_3-COOH$$

In addition, variable amounts of hydrogen are evolved. H_2 formation increases the yield of CO_2 and decreases the yield of acetate. Tracer experiments have shown that the CO_2 formed in this fermentation is derived from the carboxyl group of glycine and that both carbons of acetate originate partly from the methylene carbon of glycine and partly from CO_2. The pathway depicted in Figure 8.26 is in agreement with these results. In the first, very complex reaction glycine is cleaved and carbon 2 is transferred to tetrahydrofolate; NAD is reduced. The enzyme complex, glycine synthase, by which this reaction is carried out, consists of four proteins. Transfer of the C_1-moiety to another molecule of glycine by serine hydroxymethyltransferase yields serine. Subsequently, pyruvate is formed by serine dehydratase and

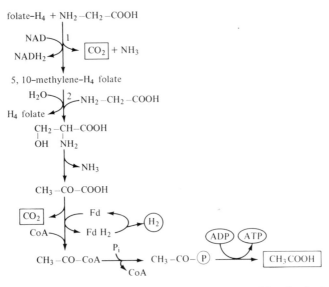

folate–H$_4$ + NH$_2$–CH$_2$–COOH

NAD ⟍ 1

NADH$_2$ ⟋ → $\boxed{CO_2}$ + NH$_3$

5, 10-methylene-H$_4$ folate

H$_2$O ⟍ 2 NH$_2$–CH$_2$–COOH

H$_4$ folate ⟋

CH$_2$–CH–COOH
| |
OH NH$_2$

→ NH$_3$

CH$_3$–CO–COOH

$\boxed{CO_2}$ ⟋ Fd

CoA ⟍ Fd H$_2$ → $\boxed{H_2}$

ADP ATP

P$_i$

CH$_3$–CO–CoA → CH$_3$–CO–$\boxed{P}$ → $\boxed{CH_3COOH}$

CoA

Figure 8.26. Fermentation of glycine by *Peptococcus anaerobius*. 1, glycine synthase enzyme complex; 2, serine hydroxymethyltransferase. Reduced ferredoxin is partly used for CO$_2$ reduction to acetate. Some of it is used for H$_2$ evolution.

finally acetate by the action of pyruvate-ferredoxin oxidoreductase, phosphotransacetylase, and acetate kinase. The NADH$_2$ formed in the first step and reduced ferredoxin are used to reduce CO$_2$ to acetate. The pathway used is probably the same as in *C. thermoaceticum* and *C. formicoaceticum*.

The fermentation of **threonine** (Figure 8.27) is initiated either by threonine dehydratase (*C. propionicum*, *Peptococcus aerogenes*) or by threonine aldolase (*C. pasteurianum*). The α-oxobutyrate formed is further converted to propionate in a reaction sequence involving an enzyme similar to pyruvate-ferredoxin oxidoreductase. *C. pasteurianum* reduces acetaldehyde to ethanol; the fate of glycine is not known.

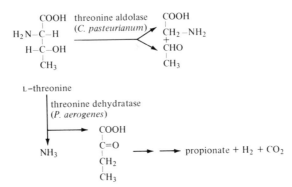

COOH threonine aldolase COOH
| (*C. pasteurianum*) |
H$_2$N–C–H ———————————→ CH$_2$–NH$_2$
| +
H–C–OH   CHO
| |
CH$_3$ CH$_3$

L–threonine

threonine dehydratase
(*P. aerogenes*)

COOH
|
NH$_3$ C=O → → propionate + H$_2$ + CO$_2$
|
CH$_2$
|
CH$_3$

Figure 8.27. Initial reactions in the fermentation of threonine.

Aspartate is fermented by many facultative and some obligate anaerobic bacteria; it is deaminated to fumarate, which is partly reduced to succinate and partly oxidized to acetate. The pathways involved are similar to those of fumarate and malate fermentation. Some clostridial species, e.g., *C. welchii*, contain a decarboxylase that converts aspartate into alanine:

$$HOOC-CH_2-CH(NH_2)-COOH \longrightarrow CO_2 + CH_3-CH(NH_2)-COOH$$

The fermentation of **glutamate** by obligate anaerobic bacteria has received considerable attention. This amino acid seems to be the preferred substrate of *Clostridium tetanomorphum*, which employs a rather unusual pathway for its breakdown. The elucidation of this pathway by Barker and collaborators led to the discovery of the first B_{12}-dependent enzyme, **glutamate mutase**. The rearrangement reaction catalyzed by this enzyme has been already discussed in connection with other B_{12}-dependent enzymes (Chapter 8, Section V). As shown in Figure 8.28(a), the product of the mutase reaction, β-methylaspartate, is deaminated to yield mesaconate. Addition of water leads to citramalate, which subsequently is cleaved to acetate and pyruvate. This reaction resembles the citrate lyase reaction, and the citramalate and citrate lyases are closely related to one another. Pyruvate is then further degraded to acetyl-CoA, CO_2, and reduced ferredoxin. Little hydrogen is evolved from the latter. Acetyl-CoA is only partly converted to acetate; the rest yields butyrate.

Peptococcus aerogenes ferments glutamate to the same products as *C. tetanomorphum* but uses another pathway. As is apparent from Figure 8.28(b), glutamate is first converted to α-hydroxyglutarate. Crotonyl-CoA is formed in a reaction not yet fully understood. The CO_2 liberated in this reaction is derived from the C_5-carboxyl group of α-hydroxyglutarate. Finally crotonyl-CoA dismutates to yield acetate and butyrate: one crotonyl-CoA is oxidized to two molecules of acetate (via acetoacetyl-CoA) and one crotonyl-CoA is reduced to butyrate (via butyryl-CoA).

Lysine is fermented by *Clostridium sticklandii* and by *Clostridium* SB_4 to acetate and butyrate. The interesting pathway employed is summarized in Figure 8.29; it involves several shifts of amino groups. Two of them, L-β-lysine mutase and D-α-lysine mutase, require coenzyme B_{12}. Depending on whether the first shift of an amino group proceeds at L- or D- lysine, different pathways are employed. The L-lysine pathway leads to acetate originating from carbon atoms 1 and 2 and butyrate originating from carbon atoms 3 to 6 of lysine. In the D-lysine pathway acetate is derived from C_5 and C_6. Clostridia normally favor the L-lysine route. Under certain conditions, however, the alternate pathway is preferred.

Arginine is fermented to ornithine, CO_2, and NH_3 by clostridia, streptococci, and mycoplasmas. This fermentation is unusual in that ATP is formed from carbamyl phosphate (Figure 8.30). Ornithine is degraded further by *C. sticklandii*. As in the fermentation of lysine, ornithine breakdown also involves a shift of an amino group.

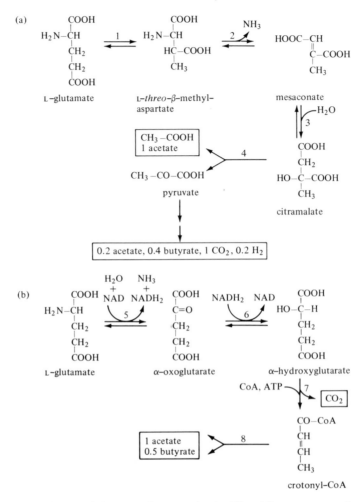

Figure 8.28. Pathways of glutamate fermentation by *Clostridium tetanomorphum* (a) and by *Peptococcus aerogenes* (b). 1, glutamate mutase; 2, β-methylaspartase; 3, citramalate dehydratase; 4, citramalate lyase; 5, glutamate dehydrogenase; 6, α-hydroxyglutarate dehydrogenase; 7, a complex reaction with glutaconyl-CoA as a probable intermediate; 8, dismutation of crotonyl-CoA.

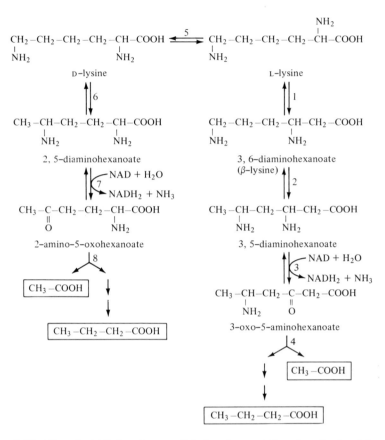

Figure 8.29. The clostridial pathway of lysine fermentation. 1, L-Lysine-2,3-amino-mutase, pyridoxal-P and Fe^{2+}-dependent, activated by S-adenosylmethionine; 2, β-lysine mutase, coenzyme B_{12}-dependent; 3, 3,5-diaminohexanoate dehydrogenase; 4, reactions not known in detail; 5, lysine racemase; 6, D-α-lysine mutase, coenzyme B_{12}-dependent; 7, 2,5-diaminohexanoate dehydrogenase; 8, reactions not known in detail.

Figure 8.30. Fermentation of L-arginine to L-ornithine. 1, arginine iminohydrolase; 2, ornithine carbamyltransferase; 3, ATP-carbamate phosphotransferase.

B. Stickland reaction

Although a number of clostridial species grow with some single amino acids, many clostridia prefer to ferment mixtures of amino acids. They carry out coupled oxidation-reductions between pairs of amino acids. One amino acid, e.g., alanine, is oxidized, and a second one, e.g., glycine, is reduced:

$$CH_3-CH(NH_2)-COOH+2H_2O \longrightarrow CH_3-COOH+CO_2+NH_3+4H$$
$$2NH_2-CH_2-COOH+4H \longrightarrow 2CH_3-COOH+2NH_3$$

This type of fermentation was discovered by Stickland in 1934; it is carried out by practically all proteolytic clostridia, such as *C. sporogenes*, *C. sticklandii*, *C. histolyticum*, and *C. botulinum*. Some amino acids are preferably used as H-donors and others as H-acceptors. The most suitable donors and acceptors are given in Table 8.9.

Amino acid oxidation proceeds via the corresponding α-oxoacid:

$$R-CH-COOH \xrightarrow{+H_2O} NH_3 + R-CO-COOH \xrightarrow{+H_2O} R-COOH+CO_2$$
$$\underset{NH_2}{|} \qquad\qquad 2H \qquad\qquad\qquad\qquad 2H$$

The first step, the oxidative deamination, is accomplished either by an enzyme of the type of glutamate dehydrogenase or by transamination with α-oxoglutarate as NH_2-acceptor and subsequent regeneration of α-oxoglutarate by glutamate dehydrogenase. The oxidative decarboxylation is catalyzed by enzymes analogous to pyruvate-ferredoxin oxidoreductase. ATP is formed from the CoA-esters by the action of CoA-transferase, phosphotransacetylase, and acetate kinase.

Amino acid reduction is a rather complex reaction. The corresponding enzyme systems are membrane-bound and consist of several proteins. The reduction of glycine to acetate requires a small selenoprotein, a large protein (protein B), and additional factors (fraction C).

Table 8.9. Amino acids that function as H-donors and as H-acceptors in Stickland reactions

H-donor	H-acceptor
alanine	glycine
leucine	proline
isoleucine	hydroxyproline
valine	ornithine
histidine	arginine
	tryptophan

$$NH_2\text{—}CH_2\text{—}COOH + NADH_2 \xrightarrow[\substack{\text{selenoprotein} \\ \text{protein B} \\ \text{fraction C}}]{} CH_3\text{—}COOH + NH_3 + NAD$$

Hydrogen is transferred from $NADH_2$ to glycine via several carriers and $NADH_2$ can be replaced in this reaction by dimercaptans such as 1,4-dithiothreitol. D-Proline reductase has similar properties. It is specific for the D-form, and clostridia carrying out Stickland reactions contain a racemase that converts L-proline into D-proline. The product of the D-proline reductase reaction is 5-aminovalerate.

$$
\begin{array}{c}
\text{CH}_2 \\
\text{H}_2\text{C} \\
\quad\quad \text{CH—COOH} + \text{NADH}_2 \longrightarrow \\
\text{H}_2\text{C} \\
\quad \text{N} \\
\quad \text{H} \\
\text{D-proline}
\end{array}
\quad
\begin{array}{c}
\text{CH}_2\text{—CH}_2\text{—CH}_2\text{—CH}_2\text{—COOH} + \text{NAD} \\
\quad\text{5-aminovalerate} \\
\text{NH}_2
\end{array}
$$

The latter compound is also formed by ornithine reductase; it is further fermented to valerate, propionate, acetate, and ammonia.

C. Heterocyclic compounds

Purines and pyrimidines are readily fermented under anaerobic conditions. *Clostridium acidi-urici* and *C. cylindrosporum* ferment guanine, hypoxanthine, urate, and xanthine; they are so specialized that they will not grow with any other substrate. The fermentation balances for urate are summarized in Table 8.10. The two clostridial species differ in that glycine is formed only by *C. cylindrosporum*.

The fermentation of purines is initiated by their conversion into xanthine (Figure 8.31). The enzyme xanthine dehydrogenase is a molybdoprotein which not only reduces urate to xanthine but converts also hypoxanthine to urate via 6,8-dihydroxypurine. The xanthine formed is then converted to formiminoglycine by a series of deaminations and decarboxylations. The latter can undergo two types of further reactions (Figure 8.32): It can be converted to glycine and formate (*C. cylindrosporum*) or it can be converted to acetate and CO_2 via serine and pyruvate (*C. acidi-urici*). The pathway leading to glycine and formate does not involve oxidations or reductions. ATP is formed from ADP and P_i by substrate-level phosphorylation in the formyltetrahydrofolate synthetase reaction. If acetate and CO_2 are the products, ATP is synthesized by acetate kinase. The hydrogen balance of this pathway is also even, because the $NADPH_2$ consumed in the methylene-tetrahydrofolate dehydrogenase reaction is regenerated from reduced ferredoxin produced by pyruvate-ferredoxin oxidoreductase. If urate is

Table 8.10. Fermentation of urate by *Clostridium acidi-urici* and *C. cylindrosporum*[a]

product	C. acidi-urici	C. cylindrosporum
	(mol of product/100 mol of urate)	
CO_2	340	275
formate	18	48
acetate	72	37
glycine	0	14
ammonia	397	356

[a]H. A. Barker, In: *The Bacteria*, I. C. Gunsalus and R. Y. Stanier (eds.). Academic Press, New York and London, 1961, vol. 2, pp. 151–207.

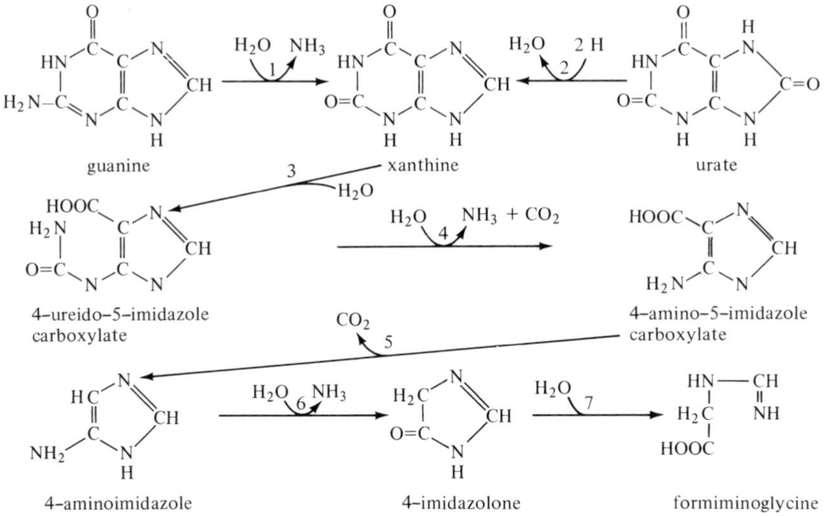

Figure 8.31. Conversion of guanine, urate, and xanthine to formiminoglycine by *C. acidi-urici* and *C. cylindrosporum*. 1, guanine deaminase; 2, xanthine dehydrogenase; 3, xanthine amidohydrolase; 4, 4-ureido-5-imidazole carboxylate amidohydrolase; 5, 4-amino-5-imidazole carboxylate decarboxylase; 6, 4-aminoimidazole deaminase; 7, 4-imidazolonase.

fermented additional reducing power is required for its reduction to xanthine. In *C. cylindrosporum* it is provided by formate dehydrogenase, which is ferredoxin-linked. In *C. acidi-urici*, part of the glycine is converted to serine by glycine synthase and serine hydroxymethyltransferase as in *Peptococcus anaerobius* (see Figure 8.26). This process is coupled to the reduction of NAD.

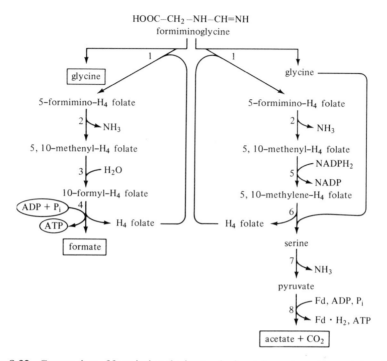

Figure 8.32. Conversion of formiminoglycine to glycine + formate and to acetate + CO$_2$. 1, glycine formiminotransferase; 2, formimino-H$_4$ folate cyclodeaminase; 3, methenyl-H$_4$ folate cyclohydrolase; 4, formyl-H$_4$ folate synthetase; 5, methylene-H$_4$ folate dehydrogenase; 6, serine hydroxymethyltransferase; 7, serine dehydratase; 8, pyruvate-ferredoxin oxidoreductase + phosphotransacetylase + acetate kinase.

A number of bacteria are known that ferment pyrimidines, e.g., *Clostridium uracilium* and *C. oroticum*. Uracil is degraded to β-alanine, CO$_2$, and NH$_3$ and orotic acid to acetate, CO$_2$, and NH$_3$. In general pyrimidines are not fermented as readily as purines.

IX. Summary

1. Fermentations are anaerobic dark processes. ATP is formed by substrate-level and/or electron transport phosphorylation.

2. Strictly anaerobic bacteria lack catalase and superoxide dismutase; they require a low redox potential for growth.

3. Yeasts ferment glucose to ethanol and CO$_2$. The key enzyme of this fermentation is pyruvate decarboxylase. Two mol of ATP are formed per mol of glucose fermented; yeasts increase the rate of glucose breakdown when transferred from aerobic to anaerobic conditions (Pasteur effect).

Zymomonas species, *Sarcina ventriculi*, and *Erwinia amylovora* carry out alcohol fermentations. Ethanol produced in smaller amounts by lactic acid bacteria, enterobacteria, and clostridia is formed by reduction of acetyl-CoA.

4. Lactic acid bacteria employ the homofermentative, the heterofermentative, or the bifidum pathway for the fermentation of hexoses. The first pathway yields two lactate/glucose, the heterofermentative pathway yields lactate, ethanol, and CO_2, whereas acetate and lactate are formed in a ratio of 3:2 by the bifidum pathway. The key enzyme of the latter two pathways is phosphoketolase.

Lactic acid is produced in several forms— D($-$), L($+$), and DL — and lactic acid bacteria contain D-lactate dehydrogenase, L-lactate dehydrogenase, or a mixture of these enzymes. A few producers of the DL-form contain L-lactate dehydrogenase plus racemase.

Lactobacillus plantarum and some other species ferment malate to lactate and CO_2 (malo-lactate fermentation). *Streptococcus cremoris* and *Leuconostoc cremoris* produce diacetyl from citrate. The latter is cleaved into acetate and oxaloacetate by citrate lyase and the final step in diacetyl synthesis is the condensation of acetyl-CoA with hydroxyethylthiamin pyrophosphate.

From growth yield studies with lactic acid bacteria it was deduced that an average of 10.5 g of cells can be formed per 1 mol ATP produced, provided that all monomers required in biosynthesis are available to the cells.

5. The main fermentation product of many clostridia, eubacteria, fusobacteria, and butyrivibrios is butyrate. It is formed from sugars via pyruvate, acetyl-CoA, acetoacetyl-CoA, and butyryl-CoA. The conversion of pyruvate to acetyl-CoA is catalyzed by pyruvate-ferredoxin oxidoreductase. In many fermentations H_2 is produced from reduced ferredoxin with hydrogenase.

Some clostridia (e.g., *C. acetobutylicum*) form acetone and butanol at pH values below 4.5. The formation of butanol is a consequence of the formation of acetone.

The hydrogen balance of fermentations can be determined on the basis of the O/R values of substrates and products or on the basis of the number of available hydrogens in substrates and products.

C. kluyveri ferments ethanol and acetate to butyrate, caproate, and molecular hydrogen. H_2 evolution is closely connected with ATP synthesis by this microorganism. Per mole H_2 evolved 0.5 mol of acetyl-CoA becomes available for ATP synthesis. Pyruvate is synthesized in *C. kluyveri* by reductive carboxylation of acetyl-CoA, and oxaloacetate by the pyruvate carboxylase reaction. Consequently, about 30% of the cellular material of *C. kluyveri* is derived from CO_2. *C. kluyveri* contains **re**-citrate synthase.

6. Microorganisms belonging to the genera *Escherichia*, *Salmonella*, and *Shigella* carry out a mixed acid fermentation and produce lactate, acetate, succinate, formate, CO_2, and H_2. Characteristic enzymes of this fermentation are pyruvate-formate lyase, which cleaves pyruvate into acetyl-CoA and formate, and formate-hydrogen lyase, which splits formate into $H_2 + CO_2$. Pyruvate-formate lyase is rapidly inactivated by oxygen.

7. Microorganisms belonging to the genera *Enterobacter*, *Serratia*, and *Erwinia* produce less acids than the above-mentioned enterobacteria but more CO_2, ethanol, and 2,3-butanediol. The first enzyme in 2,3-butanediol formation is α-acetolactate synthase.

8. *Clostridium propionicum* and *Megasphaera elsdenii* employ the acrylate pathway for the formation of propionate from lactate. Lactyl-CoA and acrylyl-CoA are intermediates of this pathway. Electron-transferring flavoprotein functions as H-carrier.

The propionibacteria and other propionate-forming microorganisms employ the succinate-propionate pathway in which succinyl-CoA and methylmalonyl-CoA function as intermediates. The interconversion of these two thioesters is catalyzed by methylmalonyl-CoA mutase, a coenzyme B_{12}-containing enzyme.

The reduction of fumarate to succinate by fumarate reductase is the only known process by which strict anaerobes gain ATP by electron transport phosphorylation. Fumarate reductase is membrane-bound and associated with menaquinone and in many organisms with a cytochrome of the b type.

9. Methanogenic bacteria ferment $CO_2 + H_2$, formate, methanol, and acetate. The reduction of CO_2 to methane proceeds via methyl-B_{12} and methylcoenzyme M. The latter compound is 2-methylmercaptoethanesulfonic acid. The fermentation leading from $CO_2 + H_2$ to methane must yield ATP by electron transport phosphorylation; the mechanism is not known.

The partial pressure of hydrogen in mud, in anaerobic digesters, and in the rumen is kept low by the action of methanogenic bacteria. This favors organisms that produce hydrogen (interspecies hydrogen transfer). Formate is not reduced directly to methane; it is oxidized first to CO_2. Methanol is a direct precursor of methane for *Methanosarcina barkeri*. With this growth substrate ATP is formed by substrate-level phosphorylation (formyl-H_4 folate synthetase reaction). Acetate is fermented to methane and CO_2 in such a way that the methyl carbon together with its attached hydrogens is converted to methane.

10. *C. formicoaceticum* and *C. thermoaceticum* ferment 1 mol of hexose to almost 3 mol of acetate. Acetate is formed by the Embden–Meyerhof pathway and by reduction of CO_2 to acetate. *Clostridium aceticum* and *Acetobacterium woodii* ferment $H_2 + CO_2$ to acetate.

11. In sulfide fermentation, the oxidation of organic compounds is coupled to the reduction of sulfate to sulfide. The substrate for reduction is adenosine-5'-phosphosulfate (APS), and reduction proceeds via sulfite, trithionate, and thiosulfate (*Desulfovibrio* species). In *Desulfotomaculum* species, sulfite is probably directly reduced to sulfide. Electron carriers such as ferredoxin, cytochrome c_3, rubredoxin, and menaquinone are involved in sulfate reduction and in ATP synthesis by electron transport phosphorylation. *Desulfuromonas acetoxidans* oxidizes acetate to CO_2 and reduces elemental sulfur to sulfide.

12. Single amino acids are fermented by a number of anaerobic bacteria:

alanine by *Clostridium propionicum*, glycine by *Peptococcus anaerobius*, threonine by *C. pasteurianum* and others, glutamate by *C. tetanomorphum*, and lysine by *C. sticklandii*. The fermentation of glutamate and lysine involves coenzyme B_{12}-containing enzymes.

Pairs of amino acids are fermented by a number of proteolytic clostridia (Stickland reaction). The oxidation of one amino acid (e.g., alanine) is coupled to the reduction of another amino acid (e.g., glycine).

Microorganisms such as *C. acidi-urici*, *C. cylindrosporum*, and *C. uracilium* are specialized for the fermentation of purine and pyrimidine bases. Urate is fermented by *C. acidi-urici* to acetate, CO_2, and ammonia.

Chapter 9
Chemolithotrophic and Phototrophic Metabolism

Chemolithotrophic and phototrophic bacteria have in common the ability to grow in mineral media, deriving their cell carbon from CO_2. The reducing power required for CO_2 reduction is obtained from inorganic compounds and energy is provided either by light-dependent reactions or by oxidation of inorganic compounds with oxygen or nitrate.

I. Chemolithotrophic Metabolism

A. Physiological groups of chemolithotrophs

As already mentioned chemolithotrophs gain energy by oxidation of an inorganic compound. Depending on the nature of the inorganic compound oxidized, five groups of chemolithotrophs are recognized; they are summarized in Table 9.1. The reactions carried out and their free energy change are also given.

Hydrogen-oxidizing bacteria have in common that they use molecular hydrogen as energy source. In other physiological properties and morphologically, however, they are very diverse. Since they all are facultative chemolithotrophs, taxonomists have preferred to classify them with their chemoheterotrophic relatives. The hydrogen-oxidizing bacteria include *Pseudomonas saccharophila, P. facilis, Alcaligenes eutrophus, Nocardia autotrophica*, and *Paracoccus denitrificans.* The latter is able to use nitrate instead of oxygen as electron acceptor.

Sulfur oxidizers are the thiobacilli, *Thiomicrospira pelophila*, a marine, spiral organism, and *Sulfolobus*, a thermophilic organism of irregular cell form. In addition, filamentous gliding organisms, such as *Beggiatoa* and

Table 9.1. Physiological groups of chemolithotrophs

group	ATP-yielding reaction	$\Delta G'_0$	
		kcal/reaction	kcal/$2e^-$
hydrogen bacteria	$H_2 + \frac{1}{2}O_2 \rightarrow H_2O$	$-56.7\ (-237.2)$	$-56.7\ (-237.2)$
sulfur bacteria	$S^{2-} + 2O_2 \rightarrow SO_4^{2-}$	$-189.9\ (-794.5)$	$-47.5\ (-198.6)$
	$S^0 + 1\frac{1}{2}O_2 + H_2O \rightarrow SO_4^{2-} + 2H^+$	$-139.8\ (-584.9)$	$-46.6\ (-195.0)$
iron bacteria	$Fe^{2+} + \frac{1}{4}O_2 + H^+ \rightarrow Fe^{3+} + \frac{1}{2}H_2O$	$-10.6^a\ (-\ 44.4)$	$-21.2^a\ (-\ 88.8)$
ammonia oxidizers	$NH_4^+ + 1\frac{1}{2}O_2 \rightarrow NO_2^- + 2H^+ + H_2O$	$-64.7\ (-270.7)$	$-21.6\ (-\ 90.2)$
nitrite oxidizers	$NO_2^- + \frac{1}{2}O_2 \rightarrow NO_3^-$	$-18.5\ (-\ 77.4)$	$-18.5\ (-\ 77.4)$

[a] $\Delta G'_0$ values for pH 0 are given; iron bacteria can grow at acidic pH values only. The $\Delta G'_0$ value for pH 7 would be -1 kcal/reaction. kJ values in parentheses.

Thiothrix are able to oxidize sulfide to elemental sulfur and subsequently to sulfate. Most sulfur bacteria are obligate chemolithotrophs. Some of them, e.g., *T. intermedius*, can grow as aerobic heterotrophs. *T. denitrificans* can utilize nitrate instead of oxygen as electron acceptor.

Thiobacillus ferrooxidans is able to use reduced sulfur compounds and ferrous ions alternatively as electron donors. The iron bacterium *Gallionella* probably also gains energy by oxidation of Fe^{2+} to Fe^{3+}. The free energy change of Fe^{2+} oxidation at low pH values is large enough to be coupled to ATP synthesis; it is rather small at neutral pH and iron bacteria cannot grow at pH values above about 4.

The **oxidation of ammonia to nitrite** is carried out by *Nitrosomonas*, *Nitrosospira*, *Nitrosovibrio*, and *Nitrosococcus* species; they all are obligate chemolithotrophs and so are the **nitrite oxidizers** *Nitrobacter*, *Nitrospina*, and *Nitrococcus*, with the exception of some *Nitrobacter* strains. Because nitrite is very toxic to most organisms, the processes of nitrite production and nitrite oxidation are remarkable.

B.   Energy production and generation of reducing power

The ratio in which H_2, O_2, and CO_2 are consumed by a growing culture of hydrogen-oxidizing bacteria is about the following:

$$4H_2 + 2O_2 \longrightarrow 4H_2O$$
$$2H_2 + CO_2 \longrightarrow \langle CH_2O \rangle + H_2O$$

$$6H_2 + 2O_2 + CO_2 \longrightarrow \langle CH_2O \rangle + 5H_2O$$

Thus, the oxidation of $4H_2$ to water yields enough ATP to allow the synthesis of cell material ($\langle CH_2O \rangle$) from CO_2 and H_2.

ATP synthesis results from oxidative phosphorylation as in aerobic respiration. Cytochromes, ubiquinone, and menaquinone have been found in membrane fractions of hydrogen-oxidizing bacteria. Differences between hydrogen-oxidizing species have been encountered as to the transfer of electrons from H_2 to the respiratory chain. *Nocardia opaca* contains a soluble hydrogenase which catalyzes the reduction of NAD by H_2. The product, $NADH_2$, then serves as H-donor for the respiratory chain. This is the exception. All other hydrogen-oxidizing bacteria studied contain a particulate hydrogenase which feeds electrons directly into the respiratory chain. This enzyme does not react with NAD.

Some hydrogen-oxidizing bacteria (*Alcaligenes eutrophus*, *Pseudomonas saccharophila*, and *P. ruhlandii*) contain, in addition to the particulate enzyme a soluble hydrogenase which reduces NAD and which is primarily responsible for the provision of $NADH_2$ for CO_2 reduction. The function of the two hydrogenases in these organisms is summarized in Figure 9.1.

A number of hydrogen-oxidizing bacteria (e.g., *Pseudomonas facilis*, *Paracoccus denitrificans*) contain a particulate hydrogenase only, and it is not

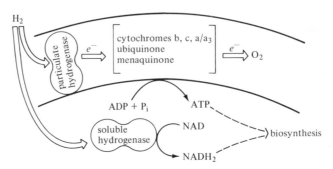

Figure 9.1. The function of the two hydrogenases of *A. eutrophus*, *P. saccharophila*, and *P. ruhlandii*. [H. G. Schlegel, *Antonie van Leeuwenhoek* **42**, 181–201 (1976).]

known how they gain $NADH_2$ for CO_2 reduction. Very likely, reverse electron transfer is involved here as it is in other chemolithotrophs.

Table 9.2 gives the redox potentials of a number of chemolithotrophic reactions. Clearly, molecular hydrogen is a good reducing agent for NAD and, as we have seen, it is used as such by a number of hydrogen-oxidizing bacteria. Out of all the other reactions only two could be coupled to NAD reduction, the oxidation of H_2S to sulfur and the oxidation of sulfite to sulfate. Experimental evidence for such reactions, however, is lacking. Compounds like ammonia or nitrite can by no means function as reducing agents for NAD; their redox potential is far too positive. Thus ammonia and nitrite oxidizers must employ **reverse electron transfer** for NAD reduction. Reverse electron transfer means that electrons are pushed up the respiratory chain from a positive to a negative redox potential at the expense of the energy of ATP hydrolysis. The situation of the nitrite oxidizers is depicted in Figure 9.2. Electron flow from nitrite to oxygen allows ATP synthesis by electron transport phosphorylation. Because of the positive redox potential of nitrite only one ATP can be formed per two electrons. A large percentage of the ATP gained in this process has to be invested into $NADH_2$ formation, from

Table 9.2. Redox potentials of reactions important in chemolithotrophic metabolism

	$E_0'(V)$
$H_2 \rightarrow 2H^+ + 2e^-$	-0.41
$NADH + H^+ \rightarrow NAD^+ + 2e^- + 2H^+$	-0.32
$H_2S \rightarrow S + 2H^+ + 2e^-$	-0.25
$S + 3H_2O \rightarrow SO_3^{2-} + 6H^+ + 4e^-$	$+0.05$
$SO_3^{2-} + H_2O \rightarrow SO_4^{2-} + 2H^+ + 2e^-$	-0.28
$NH_4^+ + 2H_2O \rightarrow NO_2^- + 8H^+ + 6e^-$	$+0.44$
$NO_2^- + H_2O \rightarrow NO_3^- + 2H^+ + 2e^-$	$+0.35$
$O_2 + 4H^+ + 4e^- \rightarrow 2H_2O$	$+0.86$

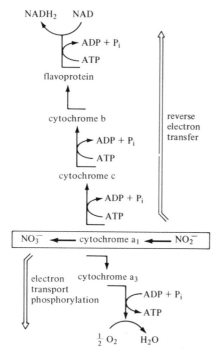

Figure 9.2. Electron transport phosphorylation and reverse electron transfer in nitrite oxidizers. Nitrite reduces cytochrome a_1 and electron transfer to oxygen is coupled to ATP synthesis. NAD is reduced by ATP-driven electron transport from cytochrome a_1 via cytochrome c, b, and flavoprotein.

NAD and nitrite. Probably more than three ATP are required per two electrons. This makes it understandable that large amounts of nitrite have to be oxidized in order to allow growth of *Nitrobacter* and that growth of these organisms is very slow.

Reverse electron transfer seems also to be involved in NAD reduction by ammonia oxidizers, iron bacteria, and sulfur bacteria. Those hydrogen-oxidizing bacteria that lack a soluble hydrogenase may also employ this mechanism for $NADH_2$ formation.

As compared to the oxidation of nitrite to nitrate, which involves one 2-electron transfer, the oxidation of ammonia and of reduced sulfur compounds is more complex. The first and the only definitely known intermediate in ammonia oxidation is **hydroxylamine**; it is formed in a monooxygenase reaction (Figure 9.3). The oxidation of hydroxylamine to nitrite is catalyzed by a particulate enzyme complex, which consists of hydroxylamine-cytochrome c reductase, and an enzyme system which oxidizes the hypothetical intermediate, **nitroxyl (NOH)**, further to nitrite. It is apparent from Figure 9.3 that only the two electrons released in the valence change of the nitrogen atom from -1 to $+1$ are fed into a respiratory chain. Whether the

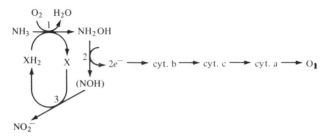

Figure 9.3. Scheme of ammonia oxidation to nitrite. 1, Monooxygenase reaction (not coupled to ATP synthesis); 2, hydroxylamine-cytochrome c reductase, which is coupled to a terminal oxidase; 3, hypothetical nitroxyl is oxidized to nitrite and regenerates XH_2, the cosubstrate in the monooxygenase reaction.

formation of one or two ATP is associated with the transport of two electrons to oxygen is not known.

Sulfur bacteria may use sulfide, elemental sulfur, or thiosulfate as reducing agents. The routes by which these compounds are oxidized to sulfate are indicated in Figure 9.4. The enzyme system which oxidizes sulfide to **elemental sulfur** is not known. However, a rapid intracellular formation of elemental sulfur by H_2S-utilizing chemolithotrophs can easily be observed. The presence of a soluble sulfur-oxidizing enzyme in several sulfur bacteria has been reported. The corresponding electron acceptors could not as yet be identified. Two enzyme systems have been found to catalyze the oxidation of **sulfite** to **sulfate**. Sulfite oxidase is membrane-bound and transfers electrons to cytochrome c. In addition, the enzyme APS reductase is present in many sulfur bacteria. Here also a cytochrome functions as electron acceptor. The adenosine-phosphosulfate (APS) formed is further converted to sulfate and ADP. The oxidation of sulfite to sulfate via APS is accompanied by substrate-level phosphorylation (AMP→ADP). Sulfur bacteria definitely gain ATP by electron transport phosphorylation; however, it is not known how much

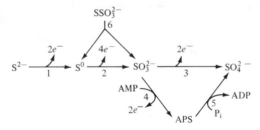

Figure 9.4. Routes of sulfur oxidation in chemolithotrophs. 1, enzyme not known; 2, sulfur-oxidizing enzyme; 3, sulfite oxidase (sulfite = cytochrome c oxidoreductase); 4, APS reductase (APS = adenosine-5′-phosphosulfate); 5, ADP sulfurylase; 6, thiosulfate-cleaving enzyme (rhodanese). [J. Suzuki, *Ann. Rev. Microbiol.* **28**, 85–101 (1974).]

they gain for the eight electrons removed from the sulfur atom in the oxidation of sulfide to sulfate.

Thiosulfate is fed into the sulfur oxidation route by cleavage into sulfite and elemental sulfur.

C. Facultative and obligate chemolithotrophs

It has been mentioned that all hydrogen-oxidizing bacteria studied so far are facultative chemolithotrophs. They grow aerobically with sugars and amino and organic acids, some of them even with purines and pyrimidines. Nevertheless, chemolithotrophic metabolism is somewhat dominant in these organisms. The induction of catabolic enzymes for organic substrate utilization is thus repressed in a H_2/O_2-atmosphere. An example is given in Figure 9.5. *Alcaligenes eutrophus* grows aerobically on fructose and employs the Entner–Doudoroff pathway for fructose degradation. Induction of the enzymes of the Entner–Doudoroff pathway is repressed in a H_2/O_2-atmosphere, and the organisms do not grow under such conditions. The mechanism of this repression is not known; it may be related to that of catabolite repression. In a H_2/O_2-atmosphere ATP is readily available, and the synthesis of catabolic enzymes could be kept repressed irrespective of the presence or absence of CO_2.

Some hydrogen-oxidizing bacteria (e.g., *Alcaligenes* Z_1) grow on formate. It serves as a quasichemolithotrophic substrate and is oxidized by formate dehydrogenase to CO_2, which is assimilated as under chemolithotrophic conditions. ATP is formed by oxidation of $NADH_2$ in the respiratory chain.

Among the nitrifiers and the sulfur bacteria facultative chemolithotrophs are rare. Some *Nitrobacter* strains grow very slowly with acetate. *Thiobacillus*

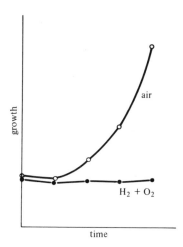

Figure 9.5. Aerobic growth of *Alcaligenes eutrophus* on fructose and repression by $H_2 + O_2$. [G. Gottschalk, *Biochem. Z.* **341**, 260–270 (1965).]

novellus and *T. intermedius* grow with organic substrates, other thiobacilli such as *T. thioparus*, *T. neapolitanus*, and *T. denitrificans* do not, and the question is, what is the basis of obligate chemolithotrophy? Several explanations can be envisaged.

1. The organisms might be impermeable to organic compounds. However, the uptake of a number of compounds and their incorporation into cellular material has been demonstrated for several obligate chemolithotrophs.

2. The formation of ATP from organic substrates might be impaired or impossible. This hypothesis has been advanced by Stanier and collaborators and, in fact, it is very plausible. Chemolithotrophs do not require a complete tricarboxylic acid cycle, and they have been shown to lack α-oxoglutarate dehydrogenase (as do the incomplete oxidizers and several strict anaerobes). Thus, $NADH_2$ for the respiratory chain cannot be generated as normally in aerobic heterotrophs. In addition, the respiratory chains of obligate chemolithotrophs might not be suited to accept electrons from $NADH_2$, as we have seen that the entry of electrons from the inorganic substrates occurs at the level of cytochromes.

3. Ammonia oxidizers and sulfur bacteria might suffer from a shortage of ammonia and reduced sulfur compounds under heterotrophic conditions. If, for instance, *Nitrosomonas* is incubated with an organic substrate in a mineral medium under air, ammonia is preferentially oxidized to nitrite and there is no N-source left for biosynthetic purposes.

The reason for obligate chemolithotrophy might not be the same in all organisms possessing this property. However, impaired energy metabolism and/or shortage of ammonia or reduced sulfur compounds offer plausible explanations.

D. Oxidation of carbon monoxide

A few microorganisms have been shown to grow aerobically on CO (e.g., Kistner's *Hydrogenomonas carboxydovorans* and Zavarzin's *Seliberia carboxyhydrogena*). CO is oxidized to CO_2 and the electrons removed are transferred to oxygen for ATP production. CO-utilizing bacteria can also oxidize molecular hydrogen, and it can be assumed that they belong to the physiological group of the hydrogen-oxidizing bacteria.

II. Assimilation of CO_2

Chemolithotrophs use the ATP and the reducing power produced by oxidation of inorganic substrates to reduce CO_2 and to convert it to cell material. Since CO_2 functions as sole source of carbon in these organisms, they are often called **C-autotrophs**. The mechanism of CO_2 fixation employed by chemolithotrophs is the Calvin cycle, the same mechanism that occurs in green plants and was discovered by Calvin, Benson, and Bassham in green algae.

A. Reactions of the Calvin cycle

The actual CO_2 fixation reaction of the Calvin cycle is the **ribulose-1,5-bisphosphate carboxylase** reaction, and the primary product of CO_2 fixation is 3-phosphoglycerate [Figure 9.6(a)]. One carboxyl group of the two molecules of 3-phosphoglycerate formed is derived from CO_2.

The next two reactions serve to reduce the carboxyl group of 3-phosphoglycerate to the aldehyde group. The enzymes involved are phosphoglycerate kinase and glyceraldehyde-3-phosphate dehydrogenase [Figure 9.6(b)].

The processes of CO_2 fixation and CO_2 reduction bring about the conversion of one molecule of ribulose-1,5-bisphosphate into two molecules of glyceraldehyde-3-phosphate. If the latter is solely used for biosyntheses, CO_2 fixation would soon come to a standstill because of a shortage of ribulose-bisphosphate. Thus a third process must follow, the regeneration of the CO_2-acceptor. It involves a whole series of reactions; most are common to us from the discussion of the glycolytic pathways and the pentose phosphate cycle (see Figures 2.2, 3.9, and 5.10). Part of the glyceraldehyde-3-

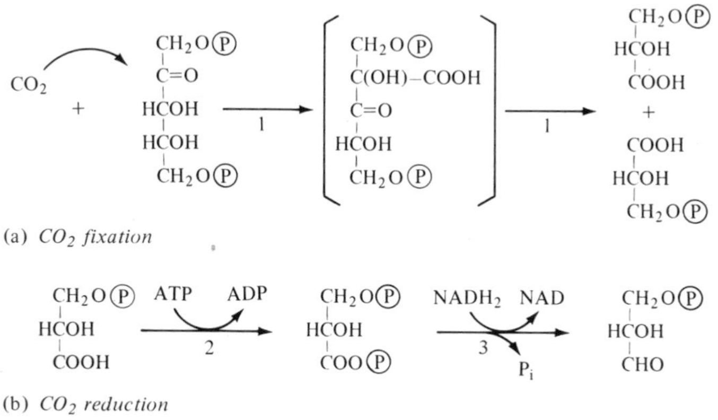

(a) *CO_2 fixation*

(b) *CO_2 reduction*

Figure 9.6. The first two phases of the Calvin cycle, CO_2 fixation (a) and CO_2 reduction (b). 1, ribulose-1,5-bisphosphate carboxylase. The unstable intermediate is 2-carboxy-3-oxoribitol-1,5-bisphosphate; 2, 3-phosphoglycerate kinase; 3, glyceraldehyde-3-phosphate dehydrogenase.

phosphate is first converted into fructose-6-phosphate by the action of triose-phosphate isomerase, fructose-1,6-bisphosphate aldolase, and phosphatase (Figure 9.7). Then, one fructose-6-phosphate, two glyceraldehyde-3-phosphates, and one dihydroxyacetone phosphate react with one another to yield 2 xylulose-5-phosphates and 1 ribose-5-phosphate.

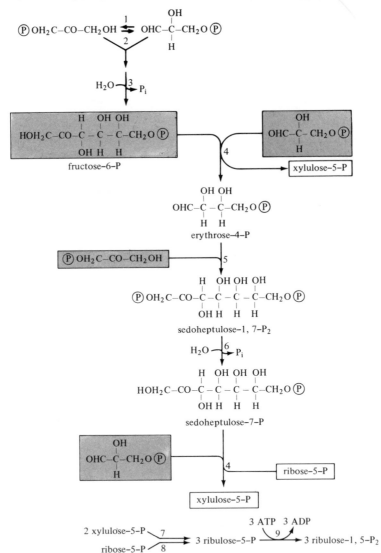

Figure 9.7. The third phase of the Calvin cycle, acceptor regeneration. 1, Triosephosphate isomerase; 2, fructose-1,6-bisphosphate aldolase; 3, fructose-1,6-bisphosphatase; 4, transketolase; 5, sedoheptulose-1,7-bisphosphate aldolase; 6, reaction is catalyzed by fructose-1,6-bisphosphatase; 7, phosphopentose epimerase; 8, ribosephosphate isomerase; 9, phosphoribulokinase.

One important difference in comparison to the pentose phosphate cycle as it occurs in heterotrophs should be pointed out. In the latter cycle a transaldolase is involved, which forms sedoheptulose-7-phosphate and glyceraldehyde-3-phosphate from erythrose-4-phosphate and fructose-6-phosphate in a reversible reaction. In the Calvin cycle, this reaction is replaced by the reversible sedoheptulose-1,7-bisphosphate aldolase and the irreversible phosphatase reaction. This guarantees a unidirectional substrate flow toward pentose phosphates, which following epimerization and isomerization reactions, are finally phosphorylated by **phosphoribulokinase** to yield the CO_2-acceptor.

The carbon balance of the whole cycle can be described schematically as follows:

$$3C_5 + \boxed{3CO_2} \longrightarrow 5C_3 + \boxed{C_3}$$
fixation and reduction
$$5C_3 \longrightarrow 3C_5$$
regeneration

The amounts of ATP and $NADH_2$ required for the formation of one glyceraldehyde-3-phosphate (GAP) from CO_2 are apparent from the following equation:

$$3CO_2 + 9ATP + 6NADH_2 \longrightarrow GAP + 9ADP + 8P_i + 6NAD$$

The whole cycle is represented in Figure 9.8.

B. Phosphoribulokinase and ribulose – 1, 5 – bisphosphate carboxylase

Two enzymes can be regarded as key enzymes of the Calvin cycle, phosphoribulokinase and ribulose-1,5-bisphosphate carboxylase. They occur only in organisms that are able to fix CO_2 via the Calvin cycle, and the evidence that the chemolithotrophs employ this cycle rests primarily on the presence of the kinase and the carboxylase in these organisms. Furthermore, it has been shown that 3-phosphoglycerate is the primary CO_2 fixation product in chemolithotrophs as it is in green plants.

Phosphoribulokinase has not been studied much. $NADH_2$ functions as positive effector for the kinase of *Pseudomonas facilis* and some other chemolithotrophs. AMP inhibits this enzyme so that ATP can be conserved in times of energy deprivation. Interestingly, PEP has been found to be a strong inhibitor of the kinase from *P. facilis*; thus an accumulation of PEP prevents the further formation of ribulose-1,5-bisphosphate, and as a consequence the further formation of phosphoglycerate.

Ribulose-1,5-bisphosphate carboxylase has been isolated from many phototrophic and chemolithotrophic organisms. The composition of this enzyme is not identical in all organisms. As is apparent from Table 9.3, three types exist. Most organisms contain a carboxylase made of eight large and

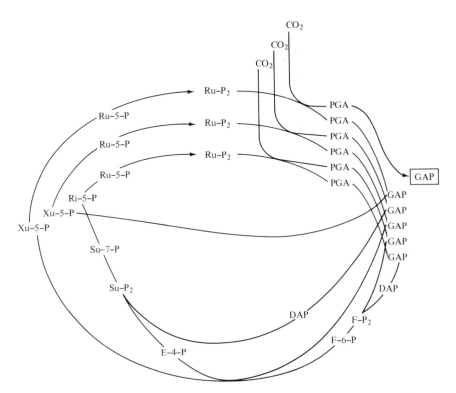

Figure 9.8. The Calvin cycle. Ru-P$_2$, ribulose-1,5-bisphosphate; PGA, 3-phospho-glycerate; GAP, glyceraldehyde-3-phosphate; DAP, dihydroxyacetone phosphate; F-P$_2$, fructose-1,6-bisphosphate; F-6-P, fructose-6-phosphate; E-4-P, erythrose-4-phosphate; Su-P$_2$, sedoheptulose-1,7-bisphosphate; Su-7-P, sedoheptulose-7-phosphate; Xu-5-P, xylulose-5-phosphate; Ri-5-P, ribose-5-phosphate; Ru-5-P, ribulose-5-phosphate.

Table 9.3. Subunit structure of ribulose-1,5-bisphosphate carboxylase from various organisms

organism	number of subunits[a]	
	large	small
Rhodospirillum rubrum	2	0
Thiobacillus intermedius	8	0
Alcaligenes eutrophus	8	8
Pseudomonas facilis	8	8
Algae and plants	8	8

[a]Approximate molecular weight of large subunit = 54,000 and of small subunit = 13,000.

eight small subunits. *Thiobacillus intermedius* carboxylase consists of eight large subunits. As compared to these types of large enzyme, the *Rhodospirillum rubrum* carboxylase is surprisingly small; it is built up of just two large subunits. This variation in composition indicates that the large subunit is the catalytic one. The function of the small subunit is unknown.

The active ribulose-1,5-bisphosphate carboxylase complex contains enzyme, Mg^{2+}, ribulose-1,5-bisphosphate, and CO_2. In the absence of CO_2 and in the presence of oxygen the carboxylase catalyzes another reaction, the **oxygenolytic cleavage of ribulose-1,5-bisphosphate** to yield phosphoglycolate and phosphoglycerate:

$$
\begin{array}{c}
CH_2O\circledP \\
| \\
C=O \\
| \\
HCOH \\
| \\
HCOH \\
| \\
CH_2O\circledP
\end{array}
\quad\xrightarrow{O_2}\quad
\begin{array}{c}
CH_2O\circledP \\
| \\
COOH
\end{array}
\quad+\quad
\begin{array}{c}
COOH \\
| \\
HCOH \\
| \\
CH_2O\circledP
\end{array}
$$

Phosphoglycolate is converted to glycolate by a phosphatase, and glycolate contributes to photorespiration in plants. Phototrophic and chemolithotrophic bacteria also produce glycolate, part of which is excreted.

Ribulose-1,5-bisphosphate carboxylase is strongly inhibited by 6-phosphogluconate. This seems reasonable for facultative chemolithotrophs. 6-Phosphogluconate signals that the oxidative pentose phosphate cycle is operating and that organic substrates are available. An active carboxylase is not required under such conditions.

III.   Phototrophic Metabolism

Photosynthesis as carried out by green plants and blue-green bacteria is accompanied by the evolution of oxygen; photosynthesis as carried out by phototrophic bacteria is different: it proceeds only under anaerobic conditions, and oxygen is not evolved. Since water cannot be utilized as hydrogen source, a suitable H-donor is required by phototrophic bacteria for growth. Such hydrogen donors are reduced sulfur compounds, molecular hydrogen, and organic substrates. Not every H-donor can be used by every phototrophic species, and on the basis of the compounds preferred and other properties three families of phototrophic bacteria are distinguished: Rhodospirillaceae, Chromatiaceae, and Chlorobiaceae.

A. The three families of phototrophic bacteria

Some properties of the phototrophic bacteria are summarized in Table 9.4. Reduced sulfur compounds are preferentially used as H-donors by the Chromatiaceae and the Chlorobiaceae. Only a few species of the Rhodospirillaceae (e.g., *Rhodopseudomonas palustris*) utilize thiosulfate and hydrogen sulfide as H-donors; none of these species can oxidize elemental sulfur. Species from all three families are able to employ H_2 for CO_2 reduction.

The relation of phototrophs to organic substrates is different. Green sulfur bacteria are able to photoassimilate only simple organic substrates such as acetate and butyrate; they will not grow in the absence of CO_2 and reduced sulfur compounds. Many purple sulfur bacteria utilize organic substrates even in the absence of reduced sulfur compounds and grow photoheterotrophically. Finally, the purple nonsulfur bacteria grow as photoheterotrophs or, in the presence of oxygen, as chemoheterotrophs, organic acids being the preferred substrates. Only one species of the sulfur bacteria grows in the dark under microaerophilic conditions, *Thiocapsa roseopersicina*. It should be stressed that anaerobic conditions are indispensable for bacterial photosynthesis. Thus, Rhodospirillaceae will grow as chemoheterotrophs in air regardless of whether light is available or not.

In addition to these metabolic differences a number of characteristic features of the three bacterial families have been recognized. The photosynthetic apparatus of the green sulfur bacteria is arranged in vesicles that are surrounded by an electron-dense membrane. The vesicles are located immediately under the cytoplasmic membrane and are readily distinguishable from it. The photosynthetic apparatus of the purple bacteria, on the other hand, is arranged in an intracytoplasmic membrane system originating from several invaginations of the cytoplasmic membrane.

The various **bacteriochlorophylls** occurring in phototrophs are listed in Table 9.4, and the chemical differences between these chlorophylls are summarized in Table 9.5. The occurrence of typical carotenoids in phototrophic bacteria has also been encountered. Green sulfur bacteria contain predominantly carotenoids bearing aromatic rings; as an example the formula of **isorenieratene** is given in Figure 9.9. A typical carotenoid of the purple bacteria is **spirilloxanthin**.

Recently, phototrophic gliding bacteria have been described. *Chloroflexus aurantiacus*, the type species, contains vesicles and resembles the green sulfur bacteria, but is able to grow as heterotroph and under aerobic conditions. Presumably, these organisms constitute a fourth family of phototrophic bacteria. The oxygenic cyanobacteria and *Prochloron* are still other sorts of phototropic prokaryotes.

Table 9.4. Some properties of the three families of phototrophic bacteria

	Rhodospirillaceae (purple nonsulfur bacteria)	Chromatiaceae (purple sulfur bacteria)	Chlorobiaceae (green bacteria)
bacteriochlorophylls present	BCHL a or b	BCHL a or b	BCHL c, d, or e
use of H_2S as H-donor	− (some +)	+	+
accumulation of S^0 and use of S^0 as H-donor	−	+	+
use of H_2 as H-donor	+	+	+
use of organic substrates as H-donor	+	+	(−)
carbon source	CO_2 organic substrates	CO_2 organic substrates	CO_2 organic substrates
aerobic dark growth	+	−	−

Table 9.5. Chemical structure of the various bacteriochlorophylls. In bacteriochlorophylls a and b carbon atoms 3 and 4 are linked by a single bond[a]

substituent	bacteriochlorophyll				
	a	b	c	d	e
R$_1$	—CO—CH$_3$	—CO—CH$_3$	—CHOH—CH$_3$	—CHOH—CH$_3$	CHOH—CH$_3$
R$_2$	—CH$_3$	—CH$_3$	—CH$_3$	—CH$_3$	—CHO
R$_3$	—C$_2$H$_5$	=CH—CH$_3$	—C$_2$H$_5$	—C$_2$H$_5$	—C$_2$H$_5$
R$_4$	—CH$_3$	—CH$_3$	—C$_2$H$_5$	—C$_2$H$_5$	—C$_2$H$_5$
R$_5$	—CO—OCH$_3$	—CO—OCH$_3$	—H	—H	—H
R$_6$	phytyl	phytyl	farsenyl	farsenyl	farsenyl
R$_7$	—H	—H	—CH$_3$	—H	—CH$_3$

[a] A. Gloe, N. Pfennig, H. Brockmann, and W. Trowitzsch, *Arch. Microbiol.* **102**, 103–109 (1975).

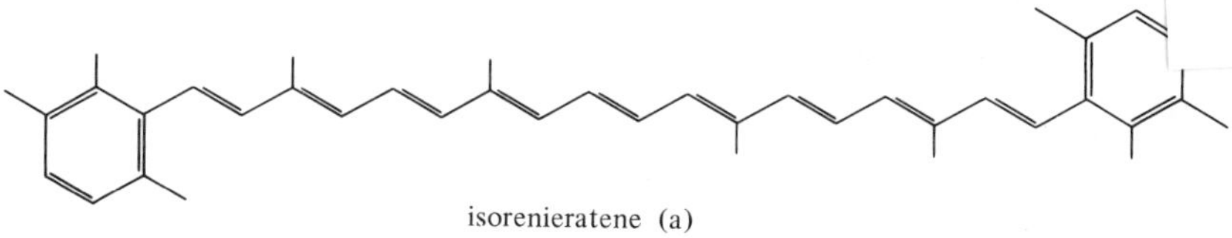

isorenieratene (a)

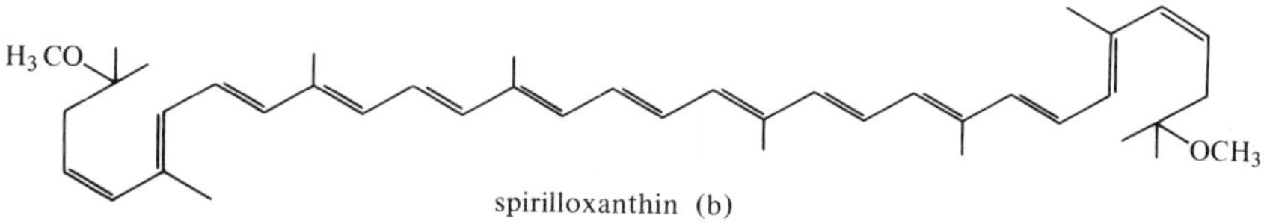

spirilloxanthin (b)

Figure 9.9. Typical carotenoids of green sulfur bacteria (a) and of purple bacteria (b).

B. Reactions of the photosynthetic apparatus

The photosynthetic apparatus contains all the components necessary for the conversion of light into ATP. It is arranged in an intracytoplasmic membrane system (purple bacteria) or in vesicles (green bacteria). Three complex reactions occur in this apparatus.

1. Light is absorbed. This is accomplished by the **antenna pigment complexes**, which contain bacteriochlorophyll and carotenoids. Often these complexes are also referred to as **light-harvesting centers**. The spectral properties of the bacteriochlorophyll and the carotenoids and their interaction with proteins determine which wavelengths are most beneficial for a particular organism. Whole cells of phototrophs exhibit two regions of absorption, one between 400 and 500 nm and one between 700 and 1,000 nm. Isolated bacteriochlorophylls absorb at shorter wavelengths (around 400 and 700 nm), and the shift to longer wavelengths *in vivo* is attributed to interactions with proteins in the antenna pigment complexes. The shift of the absorption maximum to the near infrared enables phototrophic bacteria to perform photosynthesis with light that is not absorbed by green plants and blue-green bacteria.

   Carotenoids contribute to the absorption maximum of whole cells between 400 and 500 nm. They are responsible for the red, purple, and brownish color of phototrophs and they protect cells from photooxidative damage. In addition, they are also involved in transfer of light energy.

2. Light energy causes the ejection of an electron from a bacteriochlorophyll molecule. This is accomplished in the **reaction center**. Excitation energy is channeled from the antenna pigment complexes to the reaction centers, which contain bacteriochlorophyll, bacteriopheophytin (Mg-free bacteriochlorophyll), a carotenoid, ubiquinone, and proteins. Due to strong interactions with proteins and with other components, the reaction-center bacteriochlorophyll exhibits special reactivity, and per quantum

absorbed by the center one electron is translocated from a bacteriochlorophyll to an electron transport chain:

$$\text{bacteriochlorophyll}^{\text{reactive}} + \text{light energy} \rightarrow \text{bacteriochlorophyll}^{+} + e^{\ominus}$$

The reaction center is often called **P 870** where 870 indicates the wavelength at which maximum bleaching occurs when the center loses an
electron.

3. The reaction-center bacteriochlorophyll is reduced by a high-potential
   cytochrome of the c type (e.g., cytochrome c_2). The electron generated at
   the reaction center is passed through an electron transport chain via X
   and ubiquinone to the high-potential cytochrome c. This process is
   coupled to the phosphorylation of ADP by a mechanism similar to the
   one used in the respiratory chain.

Figure 9.10 presents a scheme for the reactions proceeding in the photosynthetic apparatus. The whole process is known as **cyclic photophosphorylation**.

C. Generation of reducing power

The conversion of CO_2 to cell material in phototrophs requires, in addition
to ATP, reducing power. In plants and blue-green bacteria reducing power is
produced from water in the process of **noncyclic photophosphorylation**. This
process requires the operation of a second light reaction. A scheme is presented in Figure 9.11. Through the action of light (wavelength <680 nm), a
highly reactive chlorophyll molecule of the reaction center II is oxidized. The

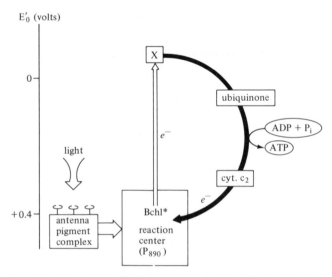

Figure 9.10. Light-dependent cyclic electron transfer and phosphorylation in phototrophic bacteria. Bchl*, Reaction center bacteriochlorophyll; X, primary acceptor.

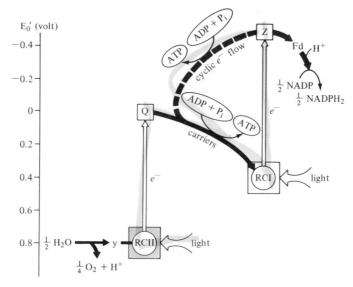

Figure 9.11. Oxygen evolution, noncyclic photophosphorylation, and NADP reduction in plants and blue-green bacteria. RC, reaction center; Q and Z, primary acceptors. The process of cyclic photophosphorylation is also indicated (RCI→Z→RCI).

electron ejected is transferred to the redox carrier Q, which has a redox potential of about zero volt. From there the electron travels via an electron transport chain to the reaction center I in order to replace an electron that has been transferred in light reaction I (wavelength $= \sim 700$ nm) to the redox system Z. The latter functions as reducing agent from ferredoxin and NADP. Transfer of electrons from Q to reaction center I is coupled to phosphorylation of ADP.

The redox potential of reaction center II is so positive that its oxidized form can be reduced by water. The latter probably reacts with an oxidized manganoprotein (Y) under evolution of oxygen, and the manganoprotein transfers electrons to the reaction center II chlorophyll.

Phototrophic bacteria do not contain the equipment necessary for performing light reaction II. Therefore, they are unable to use water as hydrogen donor. The redox potentials of the H-donors used by phototrophic bacteria are much more negative than that of water; in fact, they are usually close to the one of NAD. The E_0' values for the oxidation of sulfide to sulfur, of sulfur to sulfite, and of sulfite to sulfate are -0.25, $+0.05$, and -0.28 volts, respectively (see Table 9.2). Consequently, not much energy has to be invested in order to couple sulfide oxidation with NAD(P) reduction. Sulfide oxidation proceeds via sulfur, sulfite, and APS, as depicted in Figure 9.4, and for transfer of electrons to NAD(P) the same mechanism is probably used as in chemolithotrophs: ATP-driven **reverse electron transfer**. Thus, in contrast to oxygenic photosynthesis light seems only to be indirectly involved in the generation of reducing power; it provides the ATP by cyclic photophosphorylation. It should be mentioned that noncyclic photophosphorylation has also been discussed as a means for NAD(P) reduction in phototrophic bacteria; experimental evidence, however, is sparse.

Sulfide oxidation is strictly dependent on CO_2 fixation, and quantitative determinations of the CO_2 consumed and the sulfate produced led van Niel to formulate the following equations:

$$CO_2 + 2H_2S \xrightarrow{\text{light}} \langle CH_2O \rangle + 2S + H_2O$$

$$3CO_2 + 2S + 5H_2O \xrightarrow{\text{light}} \langle CH_2O \rangle + 2H_2SO_4$$

$$4CO_2 + 2H_2S + 4H_2O \xrightarrow{\text{light}} 4\langle CH_2O \rangle + 2H_2SO_4$$

As long as sulfide is available, it is oxidized preferentially to the level of elemental sulfur, which is deposited inside the cells (purple sulfur bacteria) or outside (green bacteria). At low sulfide concentrations the sulfur stored is oxidized further to sulfate.

A number of phototrophic bacteria can grow using molecular hydrogen as H-donor. The H_2/H^+ redox couple has a potential of $E_0' = -0.42$ V, and H_2 should be an excellent reducing agent for NAD(P). In fact, H_2 serves in this way in hydrogen -oxidizing and in methanogenic bacteria. In a number of phototrophs (*Rhodospirillum rubrum, Rhodopseudomonas capsulata*), however, the reduction of NAD by molecular hydrogen has been shown to be light-dependent. This does not mean that light is directly involved. As indicated in Figure 9.12, a high-potential cytochrome c is reduced by H_2 and ATP is required to pump the electrons from the redox potential of cytochrome c to the one of NAD. Presumably, phototrophs do not contain enzymes capable of reducing ferredoxin or NAD(P) directly with molecular hydrogen.

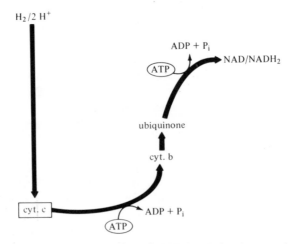

Figure 9.12. ATP-dependent reduction of NAD by H_2 in phototrophic bacteria.

Phototrophic assimilation of organic substrates is not in all cases connected with NAD reduction by reverse electron transport. During growth on ethanol, acetate, and carbohydrates $NADH_2$ is formed in the usual dehydrogenase reactions. Oxidation of succinate, however, is ATP-dependent.

$$\text{succinate} + \text{NAD} \xrightarrow{\text{ATP}} \text{fumarate} + NADH_2$$

D. Carbon metabolism

When purple bacteria grow with CO_2 as carbon source they employ the Calvin cycle for CO_2 fixation. In green sulfur bacteria, however, the key enzymes of this cycle, phosphoribulokinase and ribulose-1,5-bisphosphate carboxylase, could not be detected, and the mechanism used by these organisms for CO_2 fixation is not known. Growth of the Chlorobiaceae is stimulated by acetate. Under these conditions pyruvate is synthesized by reductive carboxylation of acetyl-CoA as in *Clostridium kluyveri*:

$$\text{acetyl-CoA} + CO_2 + FdH_2 \longrightarrow \text{pyruvate} + \text{Fd} + \text{CoA}$$

This reaction is also used for pyruvate synthesis by purple sulfur bacteria, if acetate and CO_2 are available.

Rhodospirillaceae grow on acetate anaerobically in the light and also aerobically. *Rhodospirillum tenue*, *Rhodopseudomonas palustris*, and *Rhodomicrobium vannielii* contain isocitrate lyase and malate synthase and employ the glyoxylate cycle as anaplerotic sequence. Species like *Rhodospirillum rubrum* and *Rhodopseudomonas sphaeroides* lack isocitrate lyase but contain malate synthase. How glyoxylate is formed in these organisms is not known. Nevertheless, it seems that the purple nonsulfur bacteria use a glyoxylate cycle in order to synthesize C_4-dicarboxylic acids from acetate whereas direct carboxylation reactions (acetyl-CoA→pyruvate→oxaloacetate) are employed by purple sulfur and green bacteria. Sugars are not the preferred substrates of phototrophs. Many species do not utilize them at all. *Thiocapsa roseopersicina* grows on fructose and so does *Rhodospirillum rubrum*. *Rhodopseudomonas sphaeroides* and *Rps. capsulata* utilize fructose and glucose. The former sugar is degraded via the Embden–Meyerhof pathway and the latter via the Entner–Doudoroff pathway.

Rhodopseudomonas palustris grows aerobically on *p*-hydroxybenzoate; the compound is degraded via protocatechuate and the meta-fission pathway. The key enzymes of this pathway are virtually absent from extracts of *Rps. palustris* grown phototrophically on *p*-hydroxybenzoate or benzoate. This fact led to the discovery of a new pathway for the breakdown of aromatic compounds; it does not involve O_2. As shown in Figure 9.13 benzoate is reduced to cyclohex-1-ene-1-carboxylate, which is further degraded to pimelate. How widely this pathway is distributed among phototrophs is not known.

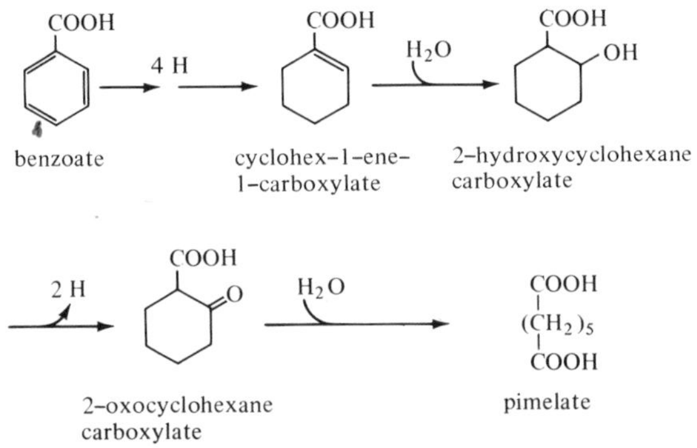

Figure 9.13. Breakdown of benzoate by *Rps. palustris* anaerobically in the light.

E. Photoproduction of molecular hydrogen

Gest and Kamen first observed that cultures of *Rhodospirillum rubrum* produced hydrogen when grown phototrophically on organic acids with amino acids (e.g., glutamate) as nitrogen source. Resting cells of such cultures photometabolize substrates such as acetate, lactate, and malate completely to CO_2 and H_2:

$$CH_3COOH + 2H_2O \xrightarrow{\text{light}} 2CO_2 + 4H_2$$

$$HOOC-CH_2-CHOH-COOH + 3H_2O \xrightarrow{\text{light}} 4CO_2 + 6H_2$$

Photoproduction of hydrogen is inhibited by NH_4^+ and by nitrogen, which indicates involvement of the enzyme nitrogenase in H_2 evolution. Like a number of other Rhodospirillaceae, *Rhodospirillum rubrum* is capable of fixing nitrogen; the formation of nitrogenase (see Chapter 10) is derepressed at low NH_4^+ concentrations (as is the case during growth with glutamate as N source), and, in the absence of molecular nitrogen, the hydrogenase activity of the nitrogenase system produces H_2 from $NADH_2$ in a reaction requiring ATP. Thus, the organic acids are oxidized via the tricarboxylic cycle, and the reduced coenzymes generated are oxidized by the nitrogenase system. Since the actual H-acceptor, N_2, is not available, the reducing power is released as molecular hydrogen.

F. Photosynthesis in halobacteria

Halobacteria live in salt lakes and are adapted to the high salt concentration in their environment and to the high light intensity in these usually shallow lakes. They are not viable at NaCl concentrations lower than 2.5 M and grow best in media containing about 5 M NaCl (sea water is about 0.6 M). The salt concentration inside the cells is the same as outside.

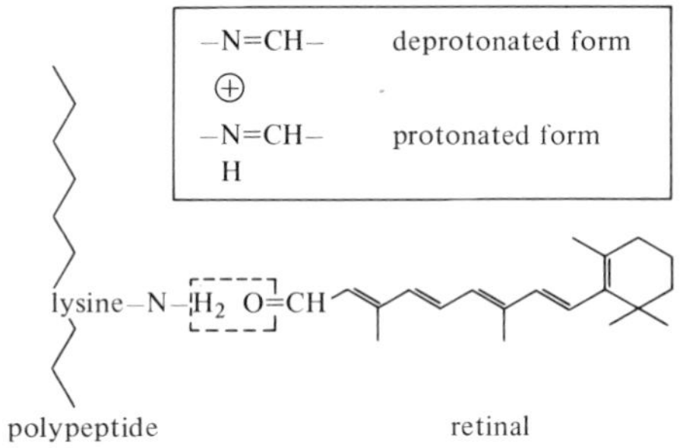

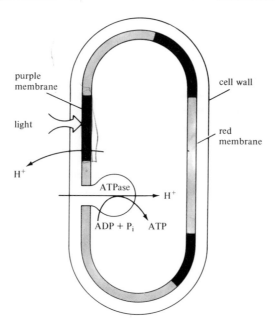

-N=CH- deprotonated form

$\oplus$

-N=CH- protonated form
H

lysine—N—[H$_2$ O=CH]/\/\/\ retinal

polypeptide   retinal

Figure 9.14. Composition of bacteriorhodopsin. Retinal, the chromophore, is linked to the amino group of a lysine residue of the polypeptide chain. The Schiff base thus formed loses and regains a proton in the photoreaction.

The membrane of the halobacteria is red and contains carotenoids which serve to prevent photochemical damage of the cells. In addition, purple-colored patches are present in the membrane. They are numerous if the cells are grown at low concentrations of oxygen in the light and they have a very interesting function. Halobacteria are obligate aerobes that utilize amino and organic acids. At low concentrations of oxygen (the solubility of O_2 in concentrated salt solutions is low), ATP is not only generated by oxidative phosphorylation but also by photosynthesis. Stoeckenius and collaborators found that the purple areas of the membrane contain **bacteriorhodopsin.** Like

Figure 9.15. Proton transport as mediated by the purple membrane in a light-powered process. [W. Stoeckenius, *Sci. Am.* **234**, 38–46 (1976).]

rhodopsin, the pigment of the rod cells of the human eye, bacteriorhodopsin consists of a protein and of the chromophore retinal, which is linked to a lysine residue of the protein (Figure 9.14). The purple membrane acts as a proton pump. It is bleached in a photoreaction, and this bleaching is accompanied by the release of a proton from bacteriorhodopsin into the medium. The process of bleaching is reversible, and the colored bacteriorhodopsin is reformed by the uptake of a proton from the cytoplasm. Thus an electrochemical proton gradient is established, which can be made use of for ATP synthesis (Figure 9.15).

IV. Summary

1. Chemolithotrophic microorganisms gain ATP by oxidation of inorganic compounds with oxygen or in some cases with nitrate. The inorganic compounds oxidized are molecular hydrogen (hydrogen-oxidizing bacteria); hydrogen sulfide, elemental sulfur, and thiosulfate (sulfur bacteria); ferrous ions (iron bacteria); ammonia (ammonia oxidizers) and nitrite (nitrite oxidizers).

2. ATP is synthesized by electron transport phosphorylation. The electrons of hydrogen are fed into a respiratory chain via a particulate hydrogenase. The oxidation of nitrite to nitrate is linked to the reduction of cytochrome a_1; its oxidation with oxygen via cytochrome a_3 yields ATP. Ammonia is oxidized to hydroxylamine in a monooxygenase reaction, and the electrons from hydroxylamine oxidation are fed into a respiratory chain. Hydrogen sulfide is oxidized to sulfate via elemental sulfur and sulfite. Two routes leading from sulfite to sulfate have been found: sulfite oxidase and APS reductase + ADP sulfurylase.

3. Cell material is formed from CO_2. With the exception of the hydrogen bacteria chemolithotrophs use reverse electron transfer in order to reduce NAD with high-potential hydrogen donors (nitrite, hydroxylamine, elemental sulfur, etc.).

4. All known hydrogen-oxidizing bacteria, *Thiobacillus novellus*, *T. intermedius* and some *Nitrobacter* strains are facultative chemolithotrophs. All the others are obligate chemolithotrophs. Reasons for obligate chemolithotrophy could be an incomplete tricarboxylic acid cycle, respiratory chains that cannot accept electrons from $NADH_2$, or a shortage of ammonia or reduced sulfur compounds for biosyntheses.

5. Chemolithotrophs use the Calvin cycle for CO_2 fixation. The key enzymes of this cycle are ribulose-1,5-bisphosphate carboxylase and phosphoribulokinase. The reduction of $3CO_2$ to glyceraldehyde-3-phosphate requires $6NADH_2$ and $9ATP$.

6. In the absence of CO_2, ribulose-1,5-bisphosphate carboxylase catalyzes an oxygenolytic cleavage of ribulose-1,5-bisphosphate into phosphoglycolate and phosphoglycerate.

7. Bacterial photosynthesis proceeds only under anaerobic conditions. Hydrogen donors used by phototrophic bacteria are reduced sulfur compounds, molecular hydrogen, and organic substrates. Rhodospirillaceae (purple nonsulfur bacteria) utilize organic substrates and H_2, Chromatiaceae (purple sulfur bacteria) utilize organic substrates, H_2, and reduced sulfur compounds, and Chlorobiaceae (green bacteria) utilize reduced sulfur compounds and H_2.

8. Phototrophic bacteria contain various bacteriochlorophylls and carotenoids. These compounds are arranged as pigment complexes in the photosynthetic apparatus. Light is absorbed by the antenna pigment complex; in the reaction center, which contains a highly reactive bacteriochlorophyll, light causes the ejection of an electron. Its passage through an electron transport chain back to the bacteriochlorophyll is coupled to ATP synthesis (cyclic photophosphorylation).

9. NAD is reduced by high-potential hydrogen donors (H_2S, S^0, succinate) using reverse electron transport.

10. Purple bacteria employ the Calvin cycle for CO_2 fixation. The mechanism of CO_2 fixation in green bacteria is unknown.

11. When grown with amino acids as nitrogen source some purple nonsulfur bacteria oxidize organic substrates to $CO_2 + H_2$. This process requires light (ATP) and involves the tricarboxylic acid cycle and the enzyme nitrogenase.

12. Halobacteria form a purple membrane when grown at low oxygen tensions. This membrane contains bacteriorhodopsin, which pumps protons outward in a light-powered reaction. The proton gradient established can be taken advantage of for ATP synthesis.

Chapter 10
Fixation of Molecular Nitrogen

That microorganisms present in soil are able to assimilate molecular nitrogen was recognized many years ago. The nitrogen gain determined by the Kjeldahl method during growth in a medium free of fixed nitrogen was usually taken as a measure of nitrogen fixation activity. In 1943 Burris and Wilson introduced the ^{15}N technique, which allowed the direct demonstration of fixed nitrogen and ammonia formation from N_2. With ^{15}N it was also possible to detect the N_2-fixing enzyme system—the **nitrogenase**—in cell extracts. In recent years a very convenient technique was introduced for the determination of nitrogenase activity, acetylene reduction. Nitrogenase reduces acetylene to ethylene, and the latter can be easily quantitated by gas chromatography.

$$N \equiv N + 6H \searrow \qquad \nearrow 2NH_3$$
$$\overline{\text{nitrogenase}}$$
$$CH \equiv CH + 2H \nearrow \qquad \searrow CH_2 = CH_2$$

This assay has contributed much to the recent progress in understanding the distribution and the mechanism of nitrogen fixation.

I.   Nitrogen-fixing Organisms

Nitrogen fixation is a property found only among prokaryotic organisms. Blue-green bacteria and bacterial species belonging to practically all known orders and families are able to fix molecular nitrogen. It is customary to distinguish between **free-living** and **symbiotic N_2-fixing organisms**. The latter exist in partnerships with plants, e.g., all members of the genus *Rhizobium* with legumes, *Klebsiella pneumoniae* with certain tropical plants, *Spirillum lipoferum* with tropical grasses, and all members of the genus *Frankia* with pioneer trees such as alder. Some N_2-fixing organisms are listed in Table 10.1.

Table 10.1. Examples of nitrogen-fixing organisms

Blue-green bacteria
  Anabaena cylindrica
  Gloeocapsa species

Phototrophic bacteria
  Rhodospirillum rubrum
  Rhodopseudomonas capsulata

Strict anaerobes
  Clostridium pasteurianum
  Desulfovibrio vulgaris

Obligate and facultative aerobes
  Rhizobium japonicum
  Frankia alni
  Klebsiella pneumoniae
  Azotobacter vinelandii
  Bacillus polymyxa
  Mycobacterium flavum
  Beijerinckia indica
  Spirillum lipoferum

II.   Biochemistry of Nitrogen Fixation

In 1960 Carnahan and collaborators announced the first successful reduction of N_2 to ammonia by a cell extract of *Clostridium pasteurianum*. The enzyme system which catalyzes this reaction is called nitrogenase. In the mean time the nitrogenase of a number of N_2-fixing organisms has been studied.

Table 10.2. Properties of the components of the nitrogenase system

property	azoferredoxin	molybdoferredoxin
molecular weight	55,000	220,000
number of subunits	2	2×2^a
iron atoms	4	18
molybdenum atoms	0	2
acid-labile sulfide	11	16

^{a} Two of each of two subunit types.

A. Composition of nitrogenase

Nitrogenase consists of two proteins in a ratio of 2:1: **azoferredoxin** and **molybdoferredoxin**. The properties of these proteins are listed in Table 10.2. Both proteins contain iron-sulfur and acid-labile sulfide. In addition, molybdoferredoxin contains two molybdenum atoms per molecule.

B. The nitrogenase reaction

The formation of ammonia from N_2 and H_2 is an exothermic reaction:

$$N_2 + 3H_2 \longrightarrow 2NH_3 \quad \Delta H = -21.8 \text{ kcal} (-91.2 \text{kJ})$$

However, N_2 is an extremely stable molecule. It contains a triple bond and the activation energy required is very high. This is why the chemical synthesis of ammonia from nitrogen and hydrogen (Haber–Bosch process) has to be carried out at high temperatures and high pressures. Nitrogenase works at room temperature. Besides nitrogen gas it requires reduced ferredoxin or flavodoxin and ATP as substrates (Figure 10.1).

Reduced ferredoxin or flavodoxin transfers electrons to the azoferredoxin. At the expense of the energy of ATP hydrolysis the potential of redox groups of the enzyme is lowered further, and finally a super-reduced molybdoferredoxin is formed, which binds N_2 and reduces it stepwise to ammonia. The binding of the dinitrogen molecule occurs probably by insertion into two metal-hydride bonds involving molybdenum.

Both subunit types of nitrogenase have ATP-binding sites, and ATP hydrolysis is associated with both the formation of the super-reductant and the reduction of the substrate. Using cell-free nitrogenase preparations the reduction of N_2 has been found to be coupled to the hydrolysis of 6 to 15 ATP depending on the conditions. *In vivo* the amount of ATP required is probably closer to 6 than to 15:

$$N_2 + 6H + (6\text{--}15) \text{ ATP} \longrightarrow 2NH_3 + (6\text{--}15) \text{ ADP} + (6\text{--}15) P_i$$

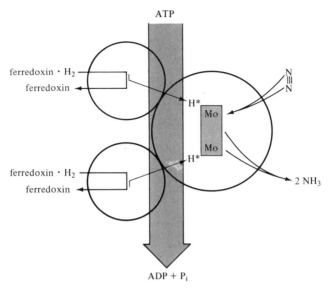

Figure 10.1. N_2 reduction to ammonia as catalyzed by nitrogenase. The small circles symbolize azoferredoxin and the large one molybdoferredoxin. H* Mo symbolizes super-reduced molybdoferredoxin.

C. Sources of reducing power

It has been mentioned that the reducing power is supplied to the nitrogenase in the form of reduced ferredoxin or flavodoxin. The former functions as reductant in strict anaerobes and phototrophs. Its reduced form is generated by the pyruvate-ferredoxin oxidoreductase or hydrogenase reactions (strict anaerobes), by noncyclic photophosphorylation (blue-green bacteria), or by reverse electron transport (phototrophic bacteria). Aerobic N_2-fixing organisms contain ferredoxin and/or flavodoxin, and these carriers are reduced with $NADH_2$ or $NADPH_2$ as H-donors.

In the absence of nitrogen the nitrogenase catalyzes an ATP-dependent evolution of molecular hydrogen.

$$5ATP + 2e^- + 2H^+ \xrightarrow{\text{nitrogenase}} H_2 + 5ADP + 5P_i$$

This reaction has been discussed in connection with the H_2 production by phototrophic bacteria (Chapter 9, Section III).

D. Oxygen sensitivity

A common property of all nitrogenase preparations is their sensitivity toward oxygen. The enzyme is irreversibly inactivated by oxygen so that nitrogen fixation can be regarded as a strictly anaerobic process. This is not surprising if one considers that a strong reductant is required for N_2-fixation. To keep the nitrogenase system anaerobic is no problem for strict anaerobes and for phototrophs. However, it is a problem for aerobes and for blue-green bacteria.

Aerobes such as *Azotobacter vinelandii* and *Mycobacterium flavum* fix nitrogen better at oxygen tensions lower than in air. For protection of nitrogenase against oxygen damage, aerobes such as the *Azotobacter* species employ two mechanisms.

1. **Respiratory protection.** *Azotobacter* species possess a very active branched respiratory system. One branch of the electron transport chain is coupled to three phosphorylation sites; the other two branches are coupled to only one phosphorylation site. Thus with increasing oxygen concentrations the rate of respiration can be increased in these organisms by a partial un-coupling. This is a waste of $NADH_2$ but it does protect the nitrogenase against oxygen damage.

2. **Conformational protection.** Following a sudden increase of the oxygen concentration, the nitrogenase of *Azotobacter chroococcum* is switched off, as shown in Figure 10.2. Nitrogenase activity appears again after lowering of the oxygen concentration. The enzyme is apparently protected by a conformational change and by the association with protective proteins.

In most filamentous blue-green bacteria the fixation of nitrogen takes place in **heterocysts**. These cells are formed when cultures are deprived of a nitrogen source other than N_2. Heterocysts are larger than vegetative cells

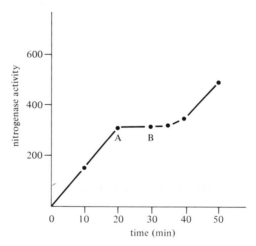

Figure 10.2. "Switch off" and "switch on" of nitrogenase. A culture of *A. chroococcum* was shaken gently under argon+oxygen (9 :1). At A, the shaking rate was increased and at B, returned to original value. [S. Hill, J. W. Drozd, and J. R. Postgate, *J. Appl. Chem. Biotechnol.* **22**, 541–548 (1972).]

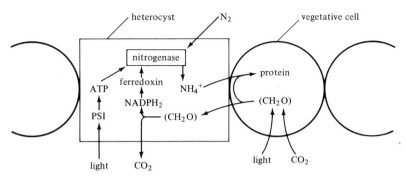

Figure 10.3. Nitrogen fixation in heterocysts. Adjacent vegetative cells provide metabolites for generation of $NADPH_2$. ATP is synthesized in the heterocysts by photophosphorylation. PSI, photosystem I.

and are surrounded by a rather thick cell wall; they are the cellular sites of nitrogen fixation. The heterocysts are devoid of photosystem II and, therefore, cannot produce oxygen. Thus, an interference of oxygenic photosynthesis with the strictly anaerobic nitrogen fixation is not possible. As illustrated in Figure 10.3 heterocysts gain ATP by photophosphorylation. Reducing power for N_2-fixation is produced from metabolites provided by the adjacent vegetative cells, which in turn receive ammonia from the heterocysts.

This very evident protective mechanism is not found in all filamentous blue-green bacteria. Some of them (e.g., *Plectonema* strains) do not form heterocysts. These species and the unicellular N_2-fixers (e.g., *Gloeocapsa* species) fix nitrogen only at low partial pressures of oxygen.

E. Symbiotic nitrogen fixation

Nitrogen fixation by the legume-*Rhizobium* symbiosis is of considerable agricultural importance. It takes place in the **root nodules**, which develop following the infection of root hairs by a *Rhizobium* species. There is a marked specificity with respect to the legume species infected; a certain *Rhizobium* strain will invade the roots of some legumes but not of others. After rhizobial infection tetraploid cells of the root are stimulated to divide, and nodules are formed in which the bacteria multiply rapidly. Finally, the bacteria are transformed into swollen, irregular cells called **bacteroids**. These bacteroids are the sites of nitrogen fixation. The path of nitrogen fixation is illustrated in Figure 10.4. N_2 is reduced by the bacteroids to NH_3, which is trapped by glutamine synthetase present in the plant cytosol. The amido group is then transferred to aspartate, and the product asparagine is removed by the root transport system. The amido group can also be transferred to α-oxoglutarate to form glutamate. In addition to asparagine the compounds glutamine and glutamate serve as important nitrogen sources to the plant. α-Oxoglutarate is generated from sucrose via the tricarboxylic acid cycle.

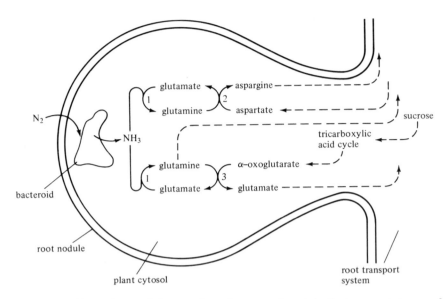

Figure 10.4. The pathway of nitrogen from the atmosphere into the transport stream of the plant. 1, glutamine synthetase; 2, glutamine=aspartate amidotransferase; 3, glutamine=α-oxoglutarate amidotransferase. [B. J. Miflin and P. J. Lea, *Trends Biochem. Sci.* **1**, 103–106 (1976).]

The plant not only provides the bacteroids with the necessary organic substrates but also shelters the nitrogenase from oxygen; N_2-fixing nodules are reddish in color because of the presence of **leghemoglobin**. This hemoglobin-like compound serves as oxygen carrier controlling access of oxygen to the bacteroids. The rhizobia are aerobes, and oxygen is transported to the bacteroids by leghemoglobin in amounts sufficient for ATP generation in the respiratory chain but low enough not to destroy the nitrogenase. Some *Rhizobium* species also fix nitrogen when grown outside their host plant in defined media.

III. Regulation of Nitrogenase

Up to 15 mol of ATP are hydrolyzed per mol of nitrogen reduced, and it is necessary for the cell to switch off nitrogen fixation under conditions of ATP starvation. This is accomplished by ADP, which is an inhibitor of nitrogenase.

The synthesis of nitrogenase is also under stringent control. It is well known that this enzyme is not synthesized if ammonia is available to the cells in sufficient amounts. Ammonia, however, not only functions as a direct repressor or co-repressor of nitrogenase synthesis. At low concentration of ammonia the level of glutamine synthetase in the cells increases considerably, the glutamine synthetase-glutamate synthase system being responsible for the incorporation of ammonia into organic compounds under these conditions (see Figure 3.2). In some microorganisms glutamine synthetase now functions as a positive control element for the synthesis of enzymes catalyzing the formation of ammonia. In *Enterobacter aerogenes* glutamine synthetase activates transcription of the histidase and proline oxidase genes. Both enzymes catalyze ammonia-yielding reactions.

The nitrogenase genes (collectively referred to as **nif** genes) of *Klebsiella pneumoniae* are under similar control; they are only transcribed if glutamine synthetase is present in the cells in a high concentration. Related control mechanisms seem to be present in other N_2-fixing microorganisms.

IV. Summary

1. Nitrogen fixation is found only among prokaryotic organisms. It occurs in strictly anaerobic, aerobic, and phototrophic organisms.

2. The enzyme which catalyzes the reduction of molecular nitrogen to ammonia is called nitrogenase. It consists of two proteins, azoferredoxin and molybdoferredoxin. For N_2-reduction, ATP and reduced ferredoxin are required as energy and hydrogen sources.

3. Nitrogenase also catalyzes the reduction of acetylene to ethylene. This reaction is used as a convenient assay of nitrogenase activity.

4. Nitrogenase is very oxygen-sensitive. Respiratory and conformational protection and heterocyst formation are used as mechanisms to prevent inactivation by oxygen. In symbiotic nitrogen fixation, the transport of oxygen to the N_2-fixing bacteroids is controlled by leghemoglobin.

5. The nitrogenase activity is inhibited by ADP. Nitrogenase is only synthesized if the ammonia concentration in the cells is low. Under these conditions the level of the enzyme glutamine synthetase is high, and glutamine synthetase functions as positive control element in the transcription of the nitrogenase genes.

Further Reading

It is recommended that standard textbooks of biochemistry and microbiology be consulted. The small selection of review articles and books given
below may be useful for further study.

Chapter 1

Guirard, B. M., Snell, E. E.: Nutritional requirements of microorganisms. In: *The
    Bacteria*, vol. 4, I. C. Gunsalus and R. Y. Stanier (eds.). Academic Press, New York,
    1962, pp. 33–95.
Hutner, S. H.: Inorganic nutrition. *Ann. Rev. Microbiol.* **26**, 313–346 (1972).

Chapter 2

Fridovich, I.: Superoxide dismutases. *Adv. Enzymol.* **41**, 35–97 (1974).
Gibson, F., Cox, G. B.: The use of mutants of *Escherichia coli* K 12 in studying electron
    transport and oxidative phosphorylation. In: *Essays in Biochemistry*, vol. 9, P. N.
    Campbell and F. Dicken (eds.). Academic Press, London and New York, 1973,
    pp. 1–29.
Harold, F. M.: Conservation and transformation of energy by bacterial membranes.
    Bacteriol. Rev. **36**, 172–230 (1972).
Kornberg, H. L.: The nature and control of carbohydrate uptake by *Escherichia coli*.
    FEBS Lett. **63**, 3–9 (1976).
Krampitz, L. O.: Cyclic mechanisms of terminal oxidation. In: *The Bacteria*, vol. 2,
    I. C. Gunsalus and R. Y. Stanier (eds.). Academic Press, New York, 1962, pp. 209–
    256.
Papa, S.: Proton translocation reactions in the respiratory chains. *Biochim. Biophys.*
    Acta **456**, 39–84 (1976).

Chapter 3

Cohen, S. N.: *Biosynthesis of Small Molecules*. Harper & Row, New York, 1967.

Costerton, J. W., Ingram, J. M., Cheng, K.-J.: Structure and function of the cell envelope of gram-negative bacteria. *Bacteriol. Rev.* **38**, 87–110 (1974).

Dagley, S., Nicholson, D. E.: *An Introduction to Metabolic Pathways*. Blackwell Scientific Publications, Oxford and Edinburgh, 1970.

Mandelstam, J., McQuillen, K.: *Biochemistry of Bacterial Growth*. Blackwell Scientific Publications, Oxford, London, Edinburgh, Melbourne, 1973.

Nikaido, H.: Biosynthesis and assembly of lipopolysaccharide and the outer membrane layer of gram-negative cell wall. In: *Bacterial Membranes and Walls*, L. Leive (ed.). Marcel Dekker, New York, 1973, pp. 131–209.

Zahner, H., Mass, W. K.: *Biology of Antibiotics*. Springer-Verlag, New York, Heidelberg, 1972.

Chapter 4

Frenkel, R.: Regulation and physiological functions of malic enzymes. *Curr. Top. Cell. Regul.* **9**, 157–182 (1975).

Kornberg, H. C.: Anaplerotic sequences and their role in metabolism. In: *Essays in Biochemistry*, vol. 2, P. N. Campbell and G. D. Greville (eds.). Academic Press, London and New York, 1966, pp. 1–32.

Chapter 5

Anderson, R. L., Wood, W. A.: Carbohydrate metabolism in microorganisms. *Ann. Rev. Microbiol.* **23**, 539–578 (1969).

Archibald, A. R.: The structure, biosynthesis and function of teichoic acid. *Adv. Microbial Physiol.* **11**, 53–90 (1974).

Dawes, E. A., Senior, P. J.: The role and regulation of energy reserve polymers in microorganisms. *Adv. Microbial Physiol.* **10**, 136–297 (1973).

Delwiche, C. C., Bryan, B. A.: Denitrification. *Ann. Rev. Microbiol.* **30**, 241–262 (1976).

Fraenkel, D. G., Vinopal, R. T.: Carbohydrate metabolism in bacteria. *Ann. Rev. Microbiol.* **27**, 69–100 (1973).

Hamilton, W. A.: Energy coupling in microbial transport. *Adv. Microbial Physiol.* **12**, 2–48 (1975).

Jones, C. W.: Aerobic respiratory systems in bacteria. In: *Microbial Energetics*, B. A. Haddock and W. A. Hamilton (eds.). Cambridge University Press, 1977, pp. 23–61.

Kaback, H. R.: Bacterial transport mechanisms. In: *Bacterial Membranes and Walls*, L. Leive (ed.). Marcel Dekker, New York, 1973, pp. 241–293.

Payne, W. J.: Denitrification. *Trends Biochem. Sci.* **1**, 220–222 (1976).

Postma, P. W., Roseman, S.: The bacterial phosphoenolpyruvate: sugar phosphotransferase system. *Biochim. Biophys. Acta* **457**, 213–257 (1976).

Reaveley, D. A., Burge, R. E.: Walls and membranes in bacteria. *Adv. Microbial Physiol.* **7**, 2–71 (1972).

Schleifer, K. H., Kandler, O.: Peptidoglycan types of bacterial cell walls and their taxonomic implications. *Bacteriol. Rev.* **36**, 407–477 (1972).

Stouthamer, A. H.: Biochemistry and genetics of nitrate reductase in bacteria. *Adv. Microbial Physiol.* **14**, 315–370 (1976).

Utter, M. F., Barden, R. E., Taylor, B. L.: Pyruvate carboxylase. *Adv. Enzymol.* **42**, 1–72 (1975).
Wood, H. G., Utter, M. F.: The role of CO_2 fixation in metabolism. In: *Essays in Biochemistry*, vol. 1, P. N. Campbell and G. D. Greville (eds.). Academic Press, London and New York, 1965, pp. 1–28.

Chapter 6

Cheldelin, V. R.: *Metabolic Pathways in Microorganisms*. John Wiley & Sons, Inc., New York and London, 1960.
Dagley, S.: Catabolism of aromatic compounds by microorganisms. *Adv. Microbial Physiol.* **6**, 1–42 (1971).
Klug, M. J., Markovetz, A. J.: Utilization of aliphatic hydrocarbons by microorganisms. *Adv. Microbiol Physiol.* **5**, 1–39 (1971).
Massey, L. K., Sokatch, J. R., Conrad, R. S.: Branched-chain amino acid catabolism in bacteria. *Bacteriol. Rev.* **40**, 42–54 (1976).
Quayle, J. R.: The metabolism of one-carbon compounds by microorganisms. *Adv. Microbial Physiol.* **7**, 119–197 (1972).
Stanier, R. Y., Ornston, L. N.: The β-ketoadipate pathway. *Adv. Microbial Physiol.* **9**, 89–149 (1973).
Takagi, T., Toda, H., Isemura, T.: Bacterial and mold amylases. In: *The Enzymes*, vol. 5, P. D. Boyer (ed.). Academic Press, New York and London, 1971, pp. 235–271.
Vogels, G. D., Van der Drift, C.: Degradation of purines and pyrimidines by microorganisms. *Bacteriol. Rev.* **40**, 403–468 (1976).
Whitaker, D. R.: Cellulases. In: *The Enzymes*, vol. 5, P. D. Boyer (ed.). Academic Press, New York and London, 1971, pp. 273–289.

Chapter 7

Cohen, G.: *The Regulation of Cell Metabolism*. Hermann, Paris, 1968.
Koshland, Jr., D. E.: Conformational aspects of enzyme regulation. *Curr. Top. Cell. Regul.* **1**, 1–28 (1969).
Ornston, L. N.: Regulation of catabolic pathways in pseudomonas. *Bacteriol. Rev.* **35**, 87–116 (1971).
Pastan, I., Adhya, S.: Cyclic adenosine-5′-monophosphate in *Escherichia coli. Bacteriol. Rev.* **40**, 527–551 (1976).
Sanwal, B. D.: Allosteric controls of amphibolic pathways in bacteria. *Bacteriol. Rev.* **34**, 20–39 (1970).
Segal, H. L.: Enzymatic interconversion of active and inactive forms of enzymes. *Science* **180**, 25–32 (1973).
Stadtman, E. R.: Mechanisms of enzyme regulation in metabolism. In: *The Enzymes*, vol. 1, P. D. Boyer (ed.). Academic Press, New York and London, 1970, pp. 398–459.
Umbarger, H. E.: Regulation of the biosynthesis of the branched chain amino acids. *Curr. Top. Cell. Regul.* **1**, 57–76 (1969).
Wyman, J.: On allosteric models. *Curr. Top. Cell. Regul.* **6**, 209–226 (1972).

Chapter 8

Barker, H. A.: *Bacterial Fermentations*. John Wiley & Sons, Inc., New York, 1956.

Barker, H. A.: Fermentation of nitrogenous organic compounds. In: *The Bacteria*, vol. 2. I. C. Gunsalus and R. Y. Stanier (eds.). Academic Press, New York, 1961, pp. 151–207.

Barker, H. A.: Coenzyme B_{12}-dependent mutases causing carbon chain rearrangements. In: *The Enzymes*, vol. 6, P. D. Boyer (ed.). Academic Press, New York and London, 1972, pp. 509–537.

Decker, K., Jungermann, K., Thauer, R. K.: Energy production in anaerobic organisms. *Angew. Chem. [Engl.]* **9**, 138–158 (1970).

Hungate, R. E.: The rumen microbial ecosystem. *Ann. Rev. Ecol. System.* **6**, 39–66 (1975).

Krebs, H. A.: The Pasteur-effect and the relations between respiration and fermentation. In: *Essays in Biochemistry*, vol. 8, P. N. Campbell and F. Dickens (eds.). Academic Press, London and New York, 1972, pp. 1–26.

Kroger, A.: Phosphorylative electron transport with fumarate and nitrate as terminal hydrogen acceptors. In: *Microbial Energetics*, B. A. Haddock and W. A. Hamilton (eds.). Cambridge University Press, 1973, pp. 61–95.

Le Gall, J., Postgate, J. R.: The physiology of sulfate-reducing bacteria. *Adv. Microbial Physiol.* **10**, 82–125 (1973).

Ljungdahl, L. G., Wood, H. G.: Total synthesis of acetate from CO_2 by heterotrophic bacteria. *Ann. Rev. Microbiol.* **23**, 515–538 (1969).

London, J.: The ecology and taxonomic status of the lactobacilli. *Ann. Rev. Microbiol.* **30**, 279–301 (1976).

Morris, J. G.: The physiology of obligate anaerobiosis. *Adv. Microbial Physiol.* **12**, 169–233 (1975).

Palmer, G.: Iron-sulfur proteins. In: *The Enzymes*, vol. 12, P. D. Boyer (ed.). Academic Press, New York and London, 1975, pp. 2–55.

Ramaiah, A.: Pasteur effect and phosphofructokinase. *Curr. Top. Cell. Regul.* **8**, 298–344 (1974).

Srere, P. A.: The enzymology of the formation and breakdown of citrate. *Adv. Enzymol.* **43**, 57–102 (1975).

Stadtman, T. C.: Lysine metabolism by clostridia. *Adv. Enzymol.* **28**, 413–447 (1973).

Wolfe, R. S.: Microbial formation of methane. *Adv. Microbial Physiol.* **6**, 107–145 (1971).

Wood, W. A.: Fermentation of carbohydrates and related compounds. In: *The Bacteria*, vol. 2, I. C. Gunsalus and R. Y. Stanier (eds.). Academic Press, New York, 1961, pp. 59–149.

Yoch, D. C., Valentine, R. C.: Ferredoxins and flavodoxins in bacteria. *Ann. Rev. Microbiol.* **26**, 139–162 (1972).

Chapter 9

Aleem, M. I. H.: Coupling of energy with electron transfer reactions in chemolithotrophic bacteria. In: *Microbial Energetics*, B. A. Haddock and W. A. Hamilton (eds.). Cambridge University Press. 1977, pp. 351–383.

Baltscheffsky, H., Baltscheffsky, M., Thore, A.: Energy conversion reactions in bacterial photosynthesis. *Curr. Top. Bioenergetics* **4**, 273–325 (1971).

Gest, H.: Energy conversion and generation of reducing power in bacterial photosynthesis. *Adv. Microbial Physiol.* **7**, 243–278 (1972).

McFadden, B. A.: Autotrophic CO_2 assimilation and the evolution of ribulose-diphosphate carboxylase. *Bacteriol. Rev.* **37**, 289–319 (1973).

Parson, W. W.: Bacterial photosynthesis. *Ann. Rev. Microbiol.* **28**, 41–59 (1974).

Pfennig, N.: Photosynthetic bacteria. *Ann Rev. Microbiol.* **21**, 285–324 (1967).

Rittenberg, S. C.: The roles of exogenous organic matter in the physiology of chemolithotrophic bacteria. *Adv. Microbial Physiol.* **3**, 159–193 (1969).

Schlegel, H. G., Eberhardt, U.: Regulatory phenomena in the metabolism of Knallgasbacteria. *Adv. Microbial Physiol.* **7**, 205–240 (1972).

Stoeckenius, W.; The purple membrane of salt-loving bacteria. *Sci. Am.* **234**, 38–58 (1976).

Suzuki, I.: Mechanisms of inorganic oxidation and energy coupling. *Ann. Rev. Microbiol.* **28**, 85–102 (1974).

Trudinger, P. A.: Assimilatory and dissimilatory metabolism of inorganic sulfur compounds by microorganisms. *Adv. Microbial Physiol.* **3**, 111–152 (1969).

Chapter 10

Benemann, J. R., Valentine, R. C.: The pathways of nitrogen fixation. *Adv. Microbial Physiol.* **8**, 59–98 (1972).

Dalton, H., Mortenson, L. E.: Dinitrogen (N_2) fixation (with a biochemical emphasis). *Bacteriol. Rev.* **36**, 231–260 (1972).

Stewart, W. D. P.: Nitrogen fixation by photosynthetic microorganisms. *Ann. Rev. Microbiol.* **27**, 283–316 (1973).

Yates. M. G., Jones, C. W.: Respiration and nitrogen fixation in Azotobacter. *Adv. Microbial Physiol.* **11**, 97–130 (1974).

Index of Organisms

Subject Index

Springer Series in Microbiology

Editor: **Mortimer P. Starr,** Department of Bacteriology, University of California, Davis, USA.

The *Springer Series in Microbiology* features textbooks and monographs designed for students and research workers in all areas of microbiology, pure and applied. Monographs by leading experts carefully summarize the state-of-the-art in various specialized aspects of microbiology. Advanced and elementary textbooks by experienced teachers treat particular microbiological topics in a manner consistent with realistic utility in classroom, laboratory, or self-instruction situations.

Thermophilic Microorganisms and Life at High Temperatures

By **T.D. Brock,** University of Wisconsin, Madison
1978. xi, 465p. 195 illus. cloth

For ten years, Professor Brock and his associates carried out an extensive research program, both in the laboratory and in the natural environment, on thermophilic microorganisms and life at high temperatures. The field work was centered in Yellowstone National Park, but additional studies were done in all the major geothermal areas of the world. The present book not only reviews this major research effort, but also covers much of the other literature on the structure, function, ecology, and practical application of thermophilic microorganisms. It will be of interest to all biologists—especially microbiologists, biochemists, ecologists, and environmental scientsists—as well as to geologists.

Microbial Ecology

Editor-in-Chief: **Ralph Mitchell,** Harvard University

Microbial Ecology, an international journal, features papers in those branches of ecology in which microorganisms are involved. Articles describe significant advances in the microbiology of natural ecosystems, as well as new methodology. In addition, the journal presents reports which explore microbiological processes associated with environmental pollution, and papers which treat the ecology of all microorganisms including pro-karyotes, eukaryotes, and viruses.

Current Microbiology

Editor: **Mortimer P. Starr,** University of California, Davis

Current Microbiology is a new journal devoted to the rapid publication of concise yet thorough research reports that deal with significant facts and ideas in all aspects of microbiology. Using an accelerated publishing technique and the active cooperation of the international editorial board, the journal is published monthly, with each issue containing from 15 to 20 brief papers. The Editorial Board and referees review each contribution at the highest professional level.

Current Microbiology presents contemporary advances in the whole field of microbiology: medical and nonmedical, basic and applied, taxonomic and historical, theoretical and practical, methodological and conceptual. The large $8\frac{1}{4}$ inch by 11 inch double-column format permits easier readability and maximum flexibility for the presentation of detailed tables and high-quality reproductions of half-tone illustrations.